WITHDRAWN
WRIGHT STATE UNIVERSITY LIBRARIES

Medical Intelligence Unit

Molecular Mechanisms of Werner's Syndrome

Michel Lebel, Ph.D.
Centre de Recherche en Cancérologie
de l' Université Laval
Hôpital Hôtel-Dieu de Québec
Québec, Canada

Landes Bioscience/Eurekah.com
Georgetown, Texas
U.S.A.

Kluwer Academic/Plenum Publishers
New York, New York
U.S.A.

MOLECULAR MECHANISMS OF WERNER'S SYNDROME

Medical Intelligence Unit

Landes Bioscience/Eurekah.com
Kluwer Academic/Plenum Publishers

Copyright ©2004 Eurekah.com and Kluwer Academic/Plenum Publishers
All rights reserved.
No part of this book may be reproduced or transmitted in any form or by any means, electronic or mechanical, including photocopy, recording, or any information storage and retrieval system, without permission in writing from the publisher.
Printed in the U.S.A.

Kluwer Academic/Plenum Publishers, 233 Spring Street, New York, New York, U.S.A. 10013
http://www.wkap.nl/

Please address all inquiries to the Publishers:
Landes Bioscience/Eurekah.com, 810 South Church Street, Georgetown, Texas, U.S.A. 78626
Phone: 512/ 863 7762; FAX: 512/ 863 0081
http://www.eurekah.com
http://www.landesbioscience.com

ISBN: 0-306-48233-9

Molecular Mechanisms of Werner's Syndrome, edited by Michel Lebel, Landes / Kluwer dual imprint / Landes series: Medical Intelligence Unit

While the authors, editors and publisher believe that drug selection and dosage and the specifications and usage of equipment and devices, as set forth in this book, are in accord with current recommendations and practice at the time of publication, they make no warranty, expressed or implied, with respect to material described in this book. In view of the ongoing research, equipment development, changes in governmental regulations and the rapid accumulation of information relating to the biomedical sciences, the reader is urged to carefully review and evaluate the information provided herein.

Library of Congress Cataloging-in-Publication Data

Molecular mechanisms of Werner's syndrome / [edited by] Michel Lebel.
  p. ; cm. -- (Medical intelligence unit)
 Includes bibliographical references and index.
 ISBN 0-306-48233-9
 1. Werner's syndrome--Molecular aspects. 2. Werner's syndrome--Genetic aspects. I. Lebel, Michel, 1967- II. Series: Medical intelligence unit (Unnumbered : 2003)
 [DNLM: 1. Werner Syndrome--genetics. 2. DNA Helicases. 3. Werner Syndrome--physiopathology. QZ 45 M718 2004]
 RC580.W47M64 2004
 616'.042--dc22
                                        2004013124

CONTENTS

Preface ix

1. **Clinical Aspects of Werner's Syndrome:**
   Its Natural History and the Genetics of the Disease 1
   Makoto Goto
   History of Werner's Syndrome Research 1
   Diagnosis of Werner's Syndrome 2
   Clinical Manifestations and Natural History of Werner's Syndrome 4
   Epidemiology 9

2. **Biochemical Roles of RecQ Helicases** 12
   Payam Mohaghegh and Ian D. Hickson
   RecQ Helicases: A Family of Molecular Motor Proteins 12
   RecQ Helicase Diseases 13
   Mouse Models of RecQ Helicase Deficiency 14
   Possible Role for WRN in DNA Replication/Repair 15
   Regulation of WRN Activity 17
   A Role for RecQ Helicases in Telomere Maintenance 17
   Biochemical Properties of RecQ Helicases 18

3. **Biochemical Characterization of the Werner Syndrome**
   DNA Helicase-Exonuclease 22
   Michael Fry
   The Human *WRN* Gene and Its Homologs 22
   Structure of the WRN Promoter 24
   Homology of WRN to Other Members of the RecQ Family 24
   The WRN Helicase Activity 24
   Tetraplex DNA 26
   Holliday Junction (4-Way X Junction) 29
   Triplex DNA 30
   Bubble-Containing Duplex DNA 30
   The WRN 3'→5' Exonuclease 32
   Coordination of the Helicase and Nuclease Activities 35
   The Transcriptional Activation Domain: WRN
      As a Transcription Activator 35
   Recent Results 36

4. **Proteins That Interact with the Werner Syndrome Gene Product** ... 44
   Dana Branzei and Takemi Enomoto
   Werner Interacting Proteins 44
   WRN Functions 55

5. **Sensitivity of Werner's Syndrome Cells to DNA Damaging Agents: Insights into the Biological Functions of the Werner Protein** 62
   Adayabalam S. Balajee and Fabrizio Palitti
   Cellular Characteristics of Werner's Syndrome 63
   Sensitivity of WS Cells to DNA Damaging Agents 64
   Role of the WRN Helicase/Exonuclease in DNA Metabolic Activities 67
   Mode of Action of the WRN Helicase/Exonuclease in Pathways
      of Genomic Stability 69

6. **Yeast RecQ Helicases: Clues to DNA Repair, Genome Stability
   and Aging** 78
   Rozalyn M. Anderson and David A. Sinclair
   Yeast RecQ Helicases: Clues to Genome Instability and Aging 78
   Phenotypes of RecQ Mutants 79
   Structures, Substrates and Localization 80
   Physical Interactions 83
   DNA Mismatch Repair (MMR) Proteins 83
   Mgs1, the Yeast WHIP Homologue 84
   Topoisomerases 84
   Rad51 86
   Rad16 86
   Genetic Interactions 86
   SLX1/4, HEX3/SLX8 and *MMS4/MUS81* 86
   SRS2 87
   Other DNA Repair Genes 88
   The Cellular Function of RecQ Helicases 88
   S Phase DNA Damage Checkpoint 88
   DNA Replication and Repair 91
   Telomere Maintenance 92
   Aging 94
   Meiosis 96

7. **Potential Function of the Werner's Syndrome Homologue
   in the African Clawed Frog and the Mouse** 107
   Michel Lebel and Philip Leder
   The Mouse and Frog *WRN* Gene Homologues 108
   Differences between Mouse Wrn and Human WRN Proteins 109
   Function of the WRN Homologue in the *Xenopus* System 110
   Function of the Wrn Homologue in Mouse 111
   Cellular Phenotype of *Wrn* Mutant Mice 111
   Phenotype of *Wrn* Mutant Mice 113
   Wrn Mutant Mice and the *p53* Tumor Suppressor Gene 115
   Illegitimate Recombination in Tissues of *Wrn* Mutant Mice 116

8. **Proposed Biological Functions for the Werner Syndrome Protein in DNA Metabolism** 123
   Patricia L. Opresko, Jeanine A. Harrigan, Wen-Hsing Cheng, Robert M. Brosh, Jr. and Vilhelm A. Bohr
   DNA Replication 123
   DNA Repair 125
   Transcription 127
   Telomere Metabolism 128

9. **Replicative Senescence, Telomeres and Werner's Syndrome** 133
   Richard G.A. Faragher
   Why Do Gerontologists Study Werner's Syndrome? 133
   The Replicative Senescence Hypothesis of Aging 134
   The Kinetics of Replicative Senescence 136
   Cell Kinetics and the Replicative Senescence Hypothesis 137
   The Potential Impact of Cellular Senescence on Human Tissues 139
   Do Cells Count Divisions and if so How? 140
   Werner's Syndrome: Putting It All Together 145

Index 153

EDITOR

Michel Lebel, Ph.D.
Centre de Recherche en Cancérologie
de l' Université Laval
Hôpital Hôtel-Dieu de Québec
Québec, Canada
Chapter 7

CONTRIBUTORS

Rozalyn M. Anderson
Department of Pathology
Harvard Medical School
Boston, Massachusetts, U.S.A.
Chapter 6

Adayabalam S. Balajee
Center for Radiological Research
Department of Radiation Oncology
College of Physicians and Surgeons
Columbia University
New York, New York, U.S.A.
Chapter 5

Vilhelm A. Bohr
Laboratory of Molecular Gerontology
National Institute on Aging
National Institutes of Health
Baltimore, Maryland, U.S.A.
Chapter 8

Dana Branzei
Molecular Cell Biology Laboratory
Graduate School of Pharmaceutical
 Sciences
Tohuku University
Aobu-ku, Sendai, Miyagi, Japan
Chapter 4

Robert M. Brosh, Jr.
Laboratory of Molecular Gerontology
National Institute on Aging
National Institutes of Health
Baltimore, Maryland, U.S.A.
Chapter 8

Wen-Hsing Cheng
Laboratory of Molecular Gerontology
National Institute on Aging
National Institutes of Health
Baltimore, Maryland, U.S.A.
Chapter 8

Takemi Enomoto
Molecular Cell Biology Laboratory
Graduate School of Pharmaceutical
 Sciences
Tohuku University
Aobu-ku, Sendai, Miyagi, Japan
Chapter 4

Richard G.A. Faragher
School of Pharmacy and Biomolecular
 Science
University of Brighton
Brighton, United Kingdom
Chapter 9

Michael Fry
Department of Biochemistry
Rappaport Faculty of Medicine
Technion-Isarel Institute of Technology
Bta Galim Haifa, Israel
Chapter 3

Makoto Goto
Department of Rheumatology
Tokyo Mteropolitan Otsuka Hospital
Tishima-ku, Tokyo, Japan
Chapter 1

Jeanine A. Harrigan
Laboratory of Molecular Gerontology
National Institute on Aging
National Institutes of Health
Baltimore, Maryland, U.S.A.
Chapter 8

Ian D. Hickson
Cancer Research UK Laboratories
Weatherall Institute of Molecular
  Medicine
University of Oxford
The John Radcliffe Hospital
Oxford, United Kingdom
Chapter 2

Philip Leder
Department of Genetics
Harvard Medical School
Boston, Massachusetts, U.S.A.
Chapter 7

Payam Mohaghegh
Cancer Research UK Laboratories
Weatherall Institute of Molecular
  Medicine
University of Oxford
The John Radcliffe Hospital
Oxford, United Kingdom
Chapter 2

Patricia L. Opresko
Laboratory of Molecular Gerontology
National Institute on Aging
National Institutes of Health
Baltimore, Maryland, U.S.A.
Chapter 8

Fabrizio Palitti
Dipartmento Di Agrobiologia E
  Agrochimica
Universita Degli Studi Della Tuscia
Viterbo, Italy
Chapter 5

David A. Sinclair
Department of Pathology
Harvard Medical School
Boston, Massachusetts, U.S.A.
Chapter 6

PREFACE

Aging is an inescapable process that all organisms undergo during the short period of time they spend on earth. Unfortunately, this highly complex process brings its share of physiological symptoms, which can have a negative impact on the quality of a particular individual's life. It is not surprising that an extensive body of research is being focused on the understanding of this mechanism, in the hopes of alleviating the manifestations of aging or, alternatively, slowing down the process. Studies on human premature aging or progeroid disorders have revealed a plethora of information that contributes to a better understanding of some aspects of aging. One such disorder is Werner's syndrome or progeria of the adult. Even though Werner's syndrome is only considered a caricature of aging by some, I remain fascinated by the simple fact that only one mutated gene (*WRN*) in humans can bring about a panoply of complex phenotypes usually associated with aging. This book was an opportunity for me to bring together the most important findings and intellectual concepts in what I can now call the field of Werner's syndrome. Apart from the first chapter, which gives an overview of the clinical manifestation of the disease, this book deals mainly with the genetics as well as the cellular and molecular aspects of the disease. My intention was not to edit the most complete book on the biology of Werner's syndrome, an impossibility considering that new findings are being recorded in the literature every other month. However, upon reading the manuscript of each contributor, I realized that, altogether, this work represents the most complete overview of what is currently known on the gene responsible for this disorder. More importantly, the reader will find a large collection of references that is essential for all biologists interested in the molecular biology of the disease.

The textbook starts with an introduction of the clinical characteristics of Werner's syndrome patients. This work would have been incomplete without first enumerating the clinical criteria for the diagnosis of Werner's syndrome. This chapter is followed by an overview of all the RecQ-type helicases, including the *WRN* gene product, known to be mutated in different human genome instability disorders. Chapter three represents a complete synopsis on the properties of the exonuclease and helicase activities of WRN protein, and is followed by a chapter on the proteins known to interact and influence WRN protein activity. The fifth chapter attempts to unravel the biological functions of WRN on the basis of Werner's syndrome cell sensitivity to different DNA damaging agents. This review, in turn, is followed by a chapter on RecQ-type helicases in yeast systems. As illustrated in this chapter, investigations of RecQ function in yeasts have produced a large body of data relevant to the function of RecQ-associated progeroid diseases. This chapter is then followed by a literature review on the potential functions of WRN homologues in African clawed frogs and mice which has contributed greatly to our understanding of the potential function of WRN during DNA replication. Chapter eight puts forward a collection of proposed biological functions for the WRN protein in several DNA metabolic activities such as DNA repair, recombination, DNA replication and gene transcription. Finally, the book wraps up with a discussion on why several gerotonlogists consider Werner's syndrome as an important disease for the study of telomere attrition, replicative senescence and aging in general.

The creation of this textbook would not have been possible without the involvement of all the authors and co-authors that have kindly accepted to contribute to this work. A large debt of gratitude should go to them. Special thanks go to R.G. Landes at Landes Bioscience and his staff (especially C. Conomos) who have greatly and efficiently assured the proper development of this book. During the editing process, I made sure that the opinions and the interpretations of the data by every author are kept intact. By doing so, I hope that this book will animate the reader to formulate novel ideas leading to creative and interesting experimentation to expand our understanding of this disease and to aging in general.

Michel Lebel, Québec

CHAPTER 1

Clinical Aspects of Werner's Syndrome:
Its Natural History and the Genetics of the Disease

Makoto Goto

Abstract

Werner's syndrome, caused by a mutation in the *WRN* (or *RecQ3*) helicase gene, shows a variety of clinical and biochemical-aging phenotypes at an early stage of life followed by death at an average age of 46 years old. It has been nominated as a top ranking progeroid syndrome. Consequently, analyses of clinical and biological deterioration of body systems observed in Werner's syndrome might shed new light on the role of gene(s) in natural human aging.

Introduction

Werner's syndrome (WS: MIM#27770) is a genetically transmitted disease with an autosomal recessive mode[1-4] and has been named *progeria adultorum* (adult form of progeria). Because of many overlaps between normal human aging and patients with this syndrome, it has been ranked the highest among natural human models of accelerated aging, or segmental form of human aging.[1,3,5] Both the rarity and the difficulty in diagnosis, especially outside of Japan, probably forced us to underestimate the incidence of WS.[3] Furthermore, despite the fact that much attention has been paid to this unique syndrome since the mapping[6] and the discovery of the WS gene (*WRN*: *RecQ3* helicase),[7] the rarity of the patients and the reduced growth potentials of their cells have minimized the study of this syndrome. Here, I will review the clinical aspects, the epidemiology, and the genetics of WS, mainly based on our 30-year survey in Japan.

History of Werner's Syndrome Research

Clarification of Disease Entity

The history of Werner's syndrome research begins with the publication of a German ophthalmologist, Otto Werner's doctoral dissertation in 1904.[8] He described several progeric manifestations in addition to skin sclerosis and bilateral juvenile cataracts in some patients. He also assumed a genetic origin of the syndrome without any evidence of their parental consanguinity. The family came from a small Alpine valley village and four siblings showed almost the same clinical signs and symptoms at a similar age. Thirty years later, two New York internists, B.S. Oppenheimer and V.H. Kugel, opened the door of Werner's syndrome research. They published two papers on the subject in 1934 and 1941 and coined the name "Werner's syndrome".[9,10] The first reference of neoplasia in WS (fibro-liposarcoma) was described by S.A. Agatson and S. Gartner in 1939.[11] A Boston internist, S.J. Thannhauser published a review article on WS in 1945[12] and Seattle-based genetists, C.J. Epstein et al, released a landmark overview in 1966.[1] They suggested the possibility of an autosomal recessive mode of inheritance for this syndrome. Thus, the clinical entity of WS was established.

Genetics

In Japan, R. Ishida, an ophthalmologist in Kyoto University, reported the first Japanese case of WS in 1917.[13] Since then, case reporting of WS has been ongoing in Japan. Autosomal recessive inheritance of WS was confirmed by the analysis of 42 Japanese families including 80 cases.[2] Seventy percent of the patients were offsprings of consanguineous marriages; mostly of first cousin-marriages. In addition, clustering of patients in the same family has been frequently reported.

The phenotype of *WRN* heterozygote has harvested a strong interest. Epstein et al[1] suggested the existence of *formes frustes*, or "abortive forms". In addition, we have reported a relatively high frequency of cancer among the family members of patients with WS.[2] However, we still do not know definitively that the heterozygous carriers show some phenotypes similar to homozygous patients.

Cellular Aging

An important paper by L. Hayflick and P. Moorhead was published in 1961.[14] They reported the limited replicative capacity of cultured skin fibroblasts and suggested a possible in vitro model of cellular aging. This paper led to the publications of G.M. Martin and his colleagues showing that the replicative potential of skin fibroblasts from healthy individuals inversely correlated with the donor age from which the skin samples were obtained.[15] Importantly, they reported the strikingly diminished cultured life-span of WS cells.[1,16]

Biochemical Research

The clinical characteristics of WS were apparent in connective tissue and include gray hair, cataracts, bird-like face, skin sclerosis, short stature with stocky trunk and extremely thin extremities. As a whole, these changes make WS patients elderly-looking.[3] M. Tokunaga and colleagues have detected excessive excretion of hyaluronic acid in the urine from WS patients and coined the term hyaluronuria,[17] which was confirmed by us.[18-20] In addition, abnormal collagen metabolism was suggested.[21] However, studies of connective tissue metabolism in WS are still rudimentary.[22]

Discovery of the Gene

In 1981, we confirmed the genetics of WS. It is transmitted as a single-gene autosomal recessive trait.[2] With the help of rapidly developing molecular biology and human genetics, we started to map the WS gene by using over 100 individuals. After a three-year study, we mapped the *WRN* gene on the short arm of chromosome 8 (8p11-12).[6] This success promptly led to the cloning of the *WRN* gene in 1996.[7] The cDNA sequence of *WRN* indicated that the central portion of the predicted protein revealed a relationship to a class of the DNA helicase family.[23] The WRN protein (also known as RecQ3) belongs to the RecQ type helicases[24] but with an additional exonuclease activity.[25] So far, 38 different mutations have been detected (personal communication with J. Oshima, Seattle). Mutations include deletions, nucleotide additions, point mutations inactivating splice acceptor sites of some of the exons and point mutations in the coding sequence of the gene. The details of the search for the *WRN* gene are reported in a chapter of a recent monograph.[26]

Diagnosis of Werner's Syndrome

Our diagnosis of WS is based on the presence of four out of five criteria when patients are under the age of 35.[2,3,27] These criteria are:
1. Consanguinity (mostly first cousins marriage)
2. Characteristic bird-like appearance and body habitus (short stature, light body weight and stocky trunk with spindly extremities)

3. Premature senescence (gray hair, alopecia, bilateral cataracts, hoarseness, osteoporosis, atherosclerosis, and malignancy)
4. Scleroderma-like skin changes (atrophic skin, skin sclerosis, skin ulcer, hyperkeratosis, hyper- or hypopigmentation, subcutaneous calcification, flat feet and telangiectasia)
5. Endocrine-metabolic disorders (diabetes mellitus, hypogonadism, thyroid dysfunction, hyper-uricemia, and hyperlipidemia)

Diagnosis of over 100 WS patients was further confirmed by more than two studies including the presence of a mutation in the *WRN* gene,[7,28-30] hyaluronuria,[18] reduced replicative life span of skin fibroblasts,[15] autoantibodies[31,32] and decreased natural killer cell activity.[33] Some clinicians use the diagnostic criteria proposed by International Registry of Werner's Syndrome group as follows (Werner Homepage: gmmartin@u.washington.edu):

A. Cardinal signs and symptoms (onset over ten years old)
   1. Cataracts (bilateral)
   2. Characteristic dermatological pathology (tight skin, atrophic skin, pigmentary alterations, ulceration, hyperkeratosis, regional subcutaneous atrophy) and characteristic facies ("bird" facies)
   3. Short stature
   4. Parental consanguinity (3rd cousin or greater) or affected sibling
   5. Premature graying and/or thinning of scalp hair
   6. Positive 24-hour urinary hyaluronic acid test (when available)
B. Additional signs and symptoms
   1. Diabetes mellitus
   2. Hypogonadism (secondary sexual underdevelopment, diminished fertility, testicular or ovarian atrophy)
   3. Osteoporosis
   4. Osteosclerosis of distal phalanges of fingers and/or toes (x-ray diagnosis)
   5. Soft tissue calcification
   6. Evidence of premature atherosclerosis (e.g., history of myocardial infarction)
   7. Mesenchymal neoplasms, rare neoplasms or multiple neoplasms
   8. Voice changes (high pitched, squeaky or hoarse voice)
   9. Flat feet

Based on these symptoms, the diagnosis of WS goes like this.

Definite WS: All the cardinal signs and two others in section B.
Probable WS: The first three cardinal signs and two others in B.
Possible WS: Either cataracts or dermatological alterations and any four others in section B.
Exclusion: Onset of signs and symptoms before adolescence (except stature, since current data on pre-adolescent growth patterns are inadequate).

Unfortunately, we do not know the specificity and the sensitivity of the criteria described above for clinically diagnosing WS. Moreover, since patients with WS show a wide variety of clinical manifestations, case reporting has been done from virtually all areas of medicine. These include internal medicine (diabetes mellitus and atherosclerosis), surgery (various cancers), neurosurgery (meningioma), psychiatry (schizophrenia), gynecology and urology (hypogonadism), ophthalmology (cataracts), dermatology (skin sclerosis and melanoma), oto-rhino-laryngology (hoarseness), plastic surgery (skin ulcer), radiology (subcutaneous calcification) and orthopedic surgery (leg gangrene and osteoporosis). So, the depth and width of the clinical descriptions of the patients varied depending upon the doctors' specialty and interest. Also, information of the signs and symptoms observed in the patients was often subjective, retrospective and subject to error. With the recent improvement of modern clinical laboratory techniques, a variety of clinical and laboratory examination is available to detect subtle physiologic changes. This will help us to diagnose a rare disease like WS more easily than before.

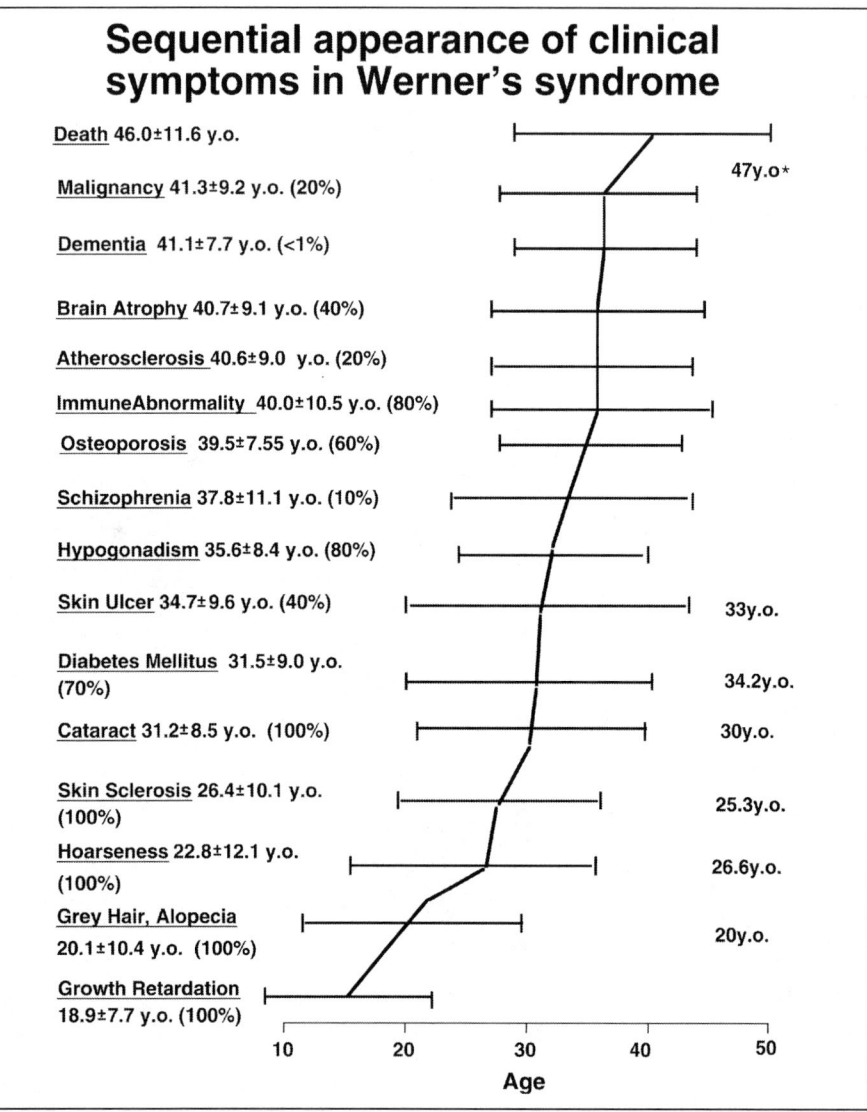

Figure 1. Sequential appearance of clinical symptoms in Werner's syndrome. The average age (± SD) at which the typical clinical manifestations were observed in Werner's syndrome is depicted. The data from Epstein et al. for comparison with patients outside Japan (mostly Caucasians and very few Blacks), is also included in the figure and is labeled with an asterisk (*).

Clinical Manifestations and Natural History of Werner's Syndrome

The sequentially appearing clinical hallmarks of patients with WS are shown in Figure 1. After a relatively normal infancy, they manifest a failure of growth spurt at age 18, followed by a hierarchical deterioration of the four major self-assembly (self-organization) body-systems as summarized in Table 1 and Figure 2. Death occurs on average at 46 years old.[1,3] The major causes of death are neoplasia and myocardial infarction. Although there is not enough data on

Table 1. Major self-assembly body systems in Werner's syndrome

Self-Assembly Body System	Mediators Inside and Between Systems	Symptoms
Connective tissue	Adhesion molecules (glycosaminoglycans, fibronectin, collagens) Cytokines	Fibrosis, cataract, skin sclerosis, osteoporosis, gray hair, arthropathy
Immune	Hormones (thyroxine, insulin, sex hormones) Metabolites	Diabetes, hypogonadism, Graves' disease Hyperlipidemia, gout
Endocrine-metabolic	Cytokines, NK cells Autoantibodies	SLE, Sjogren's, Graves' diseases
Nervous-psychological	Neuropeptides? Cytokines	Brain atrophy, schizophrenia
Mixed		Malignancy, atherosclerosis

Mediators which may affect the deterioration process are indicated. Typical symptoms in the respective body systems are also shown. SLE= Systemic Lupus Erythematosus.

normal aging to compare with those observed in WS, informations on normal development of the self-assembly body system from birth to late maturity with a similar classification was published in 1930.[34]

Connective Tissue System Disorders

The parents of children with WS usually recognize the abnormality by their lack of the prepubertal growth spurt. Basically, all WS patients show the following clinical signs and symptoms before the age of 32: characteristic habitus including slender extremities with stocky trunk (Cushingoid appearance), short stature (due to an early closure of bone end plate (range: 122-161cm) and light body weight (range: 19-52kg) (Fig. 3), bird-like appearance with pinched nose and atrophic auricle (Fig. 4), gray hair or alopecia, scleroderma-like skin changes including atrophic skin, atrophic subcutaneous tissue and muscle, hyper-or hypo-pigmentation, circumscribed hyperkeratosis, tight skin over the bones of the feet and telangiectasia, bilateral cataracts, weak and high-pitched voice (hoarseness). Both skin ulcers and subcutaneous calcification, which are not usually associated with normal aging or autoimmune scleroderma, are found in 60-80% of the WS patients. Osteoporosis, either peripheral or vertebral, is observed in over 60% of patients.[35,36] The clinical manifestations described above may be mediated by connective tissue metabolites (adhesion molecules) such as gycosaminoglycans, fibronectin and collagens (Table 1).

Endocrine-Metabolic System Disorders

Eighty percent of the WS patients before age 36 are recognized as having at least one of the following clinical signs and symptoms. Hypogonadism is observed in 80% of the patients, but about half of them show signs of hypogonadism after age 30 and have offsprings (secondary hypogonadism).[3] Non-insulin dependent diabetes mellitus specific type is associated with 70% of WS at age 36.[3] The mechanisms by which these clinical manifestations are induced are still unclear. However, in vitro experiments suggest an insulin-resistant mechanism: loss of signal transduction after its binding to normal insulin receptors.[37] Abnormality of the thyroid gland

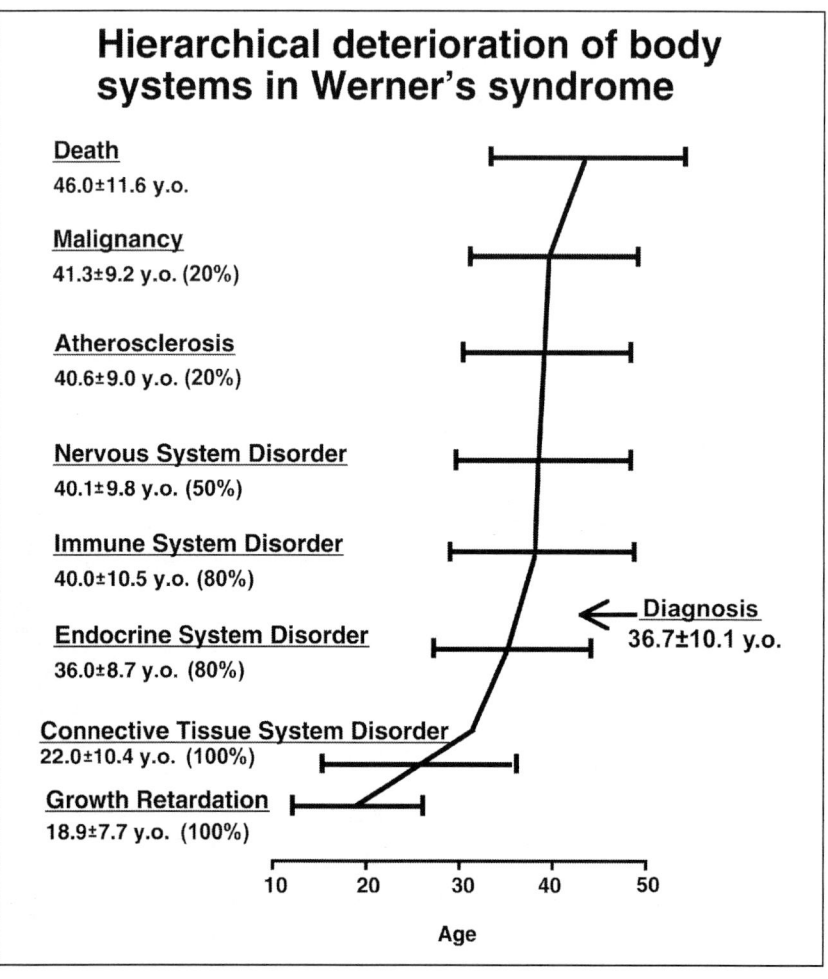

Figure 2. Hierarchical deterioration of body systems in Werner's syndrome. Relative percentage of the affected body system at the indicated age is shown.

(~15%) includes Graves' disease, hypothyroidism and thyroid carcinoma. Although, hyperuricemia is not usually associated with healthy elderly individuals, all types of hyperuricemia are found in WS: hyposecretion of uric acid, hyperproduction of uric acid and a combination of the two types.[38] Finally, hyperlipidemia characterized by hypertriglyceridemia is a biochemical hallmark of WS.[3,39]

Immune System Disorders

The immune system is a very sensitive system during normal aging.[40,41] Before the age of 40, 80% of WS patients show signs of mild immune abnormality. Deficiency in the T cell subset reactive against anti-brain associated antigens was found in all patients examined.[31] However, the number of patients studied was limited and the nature of the T cell subset remained undefined. Decreased NK cell activity, which could be recovered after interferon stimulation, was observed in most of the patients examined.[33] Most patients had low titers of several

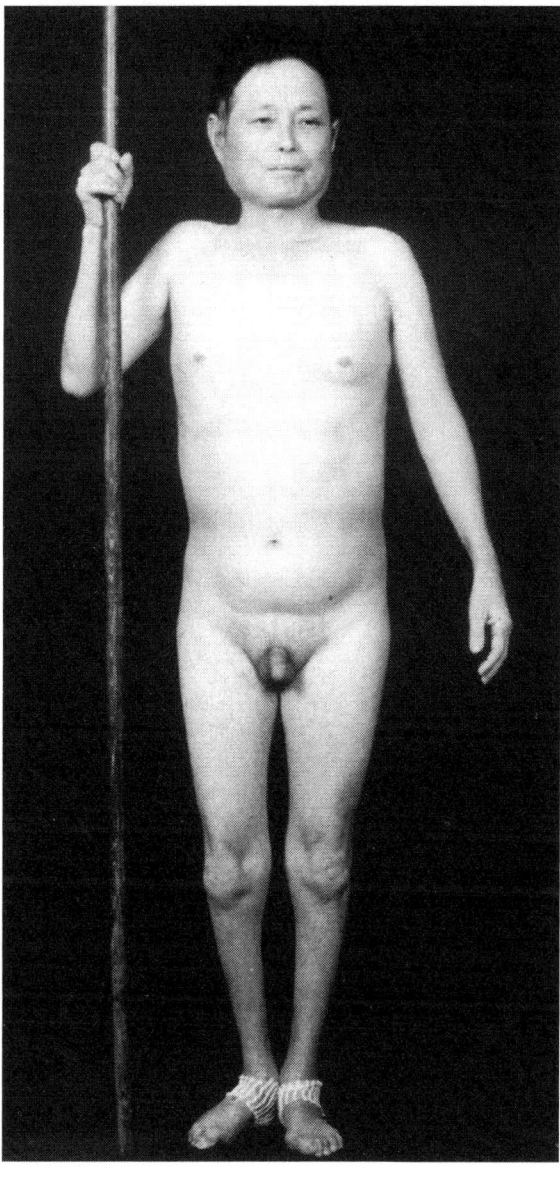

Figure 3. A 40-year-old male patient showing a typical Cushingoid appearance (a stocky trunk with extremely thin extremities). This patient was totally bald and was wearing an artificial wig. The subject also had gynecomastia, testicular atrophy and skin ulcers on both heels. This patient died of renal insufficiency derived from severe atherosclerosis of renal arteries at age 42.

autoantibodies including anti-double stranded DNA antibody, anti-nuclear antibody and rheumatoid factor, as it is usually observed in the healthy population over age 60.[31,32] The autoantibody specific to autoimmune systemic sclerosis, anti-topoisomerase I (Scl 70), has never been detected in WS, though anti-nucleolar antibody was observed in some cases.[42] Interestingly, a small percentage of patients had autoimmune diseases including Graves' disease, systemic lupus erythematosus and Sjogren's syndrome.[40] Remarkably, WS patients were not abnormally sensitive to bacterial or viral infection at any stage of their life, even though the third cause of death in WS is bacterial pneumonitis.

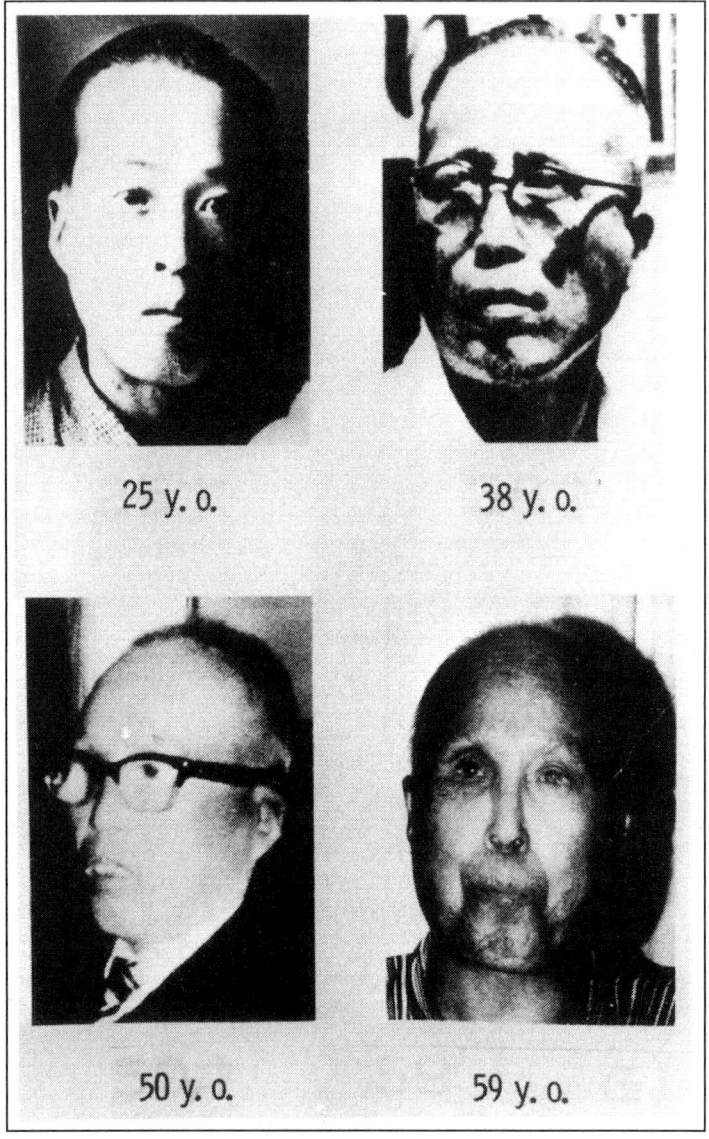

Figure 4. Typical face of a male patient at different ages. This patient died of thyroid cancer at age 60.

Nervous System Disorders

WS patients were believed to have a relatively normal central nervous system.[43] However, with the recent advance of medical devices including computed tomography and magnetic resonance imaging, brain atrophy has been observed in 40% of the WS patients even before the age of 40.[3,44] At least three patients with WS have been diagnosed as having senile dementia, but not of the Alzheimer type by clinical determination and autopsy. It is interesting to note that 10% of the patients had schizophrenia at the age of 37 as shown in Figure 1. Parkinsonism is closely associated with normal aging, but this is not the case with WS. Both far-sightedness and hearing loss are major normal aging phenotypes. Ten percent of WS patients reported before 1970 showed signs of hearing loss due to the result of otitis media infection. We do not know the far-sightedness status of the patients probably because of their cataract operation.

Mixed System Disorders

This type of self-assembly body system phenotype is characterized by at least two of the body system disorders described above and the end-results of normal aging. Atherosclerosis and malignancy, the two major causes of death in WS, may be included in this category. Atherosclerosis-related diseases including myocardial infarction, angina pectoris and hypertension were observed in 50% of the WS patients before the age of 40. This consists mainly of decreased elasticity of blood vessels based on the increased uptake of abnormal lipoproteins by macrophages. At least three systems (connective tissue, endocrine-metabolic and immune systems) participate in this process. The incidence of malignancy, especially of mesenchymal origin (sarcoma), is peculiarly high in WS (20%) before the age of 41.[45] The high incidence of malignancy in WS could be explained by its genetic instability based on the mutation of RecQ3 (*WRN*) helicase gene. D. Salk described this genetic instability as variegated translocation mosaicism[22] and as a high frequency of somatic mutations.[46,47]

Epidemiology

Since the first description of WS by Otto Werner in 1904, case reporting has been accumulated up to ~1270 worldwide in early 2002.[3] About 80% of them are of Japanese ancestry. However, no patient has ever been reported in most of the Asian countries including the mainland of China and Korea, in spite of a historically frequent racial-exchange with Japan. No patient has ever been reported from most of the African countries. Some patients reported in the United States were Japanese-American.[1] The frequency of WS was roughly estimated to be 1:100,000 in Japan and 1:1,000,000 to 1:10,000,000 outside of Japan. The only exception is the clustering of WS in Sardinia, Italy.[48] The discovery of the gene (*WRN*) enabled us to calculate the frequency of heterozygous carrier which is 1:100 in the Japanese population.[4] Mutation 4, the most frequent mutation in Japanese patients, is found in ~60% of the WS patients.[49,50] This mutation is a G to C substitution destroying the splice acceptor site of exon 26 of the *WRN* gene. Finally, in Japan, several clustering areas have been noted including Miyagi Prefecture, South Kanto District, Ishikawa Prefecture, Hyogo Prefecture, and South Kyusyu District.[2,3,49,51]

Conclusion and Future Perspectives

Progeroid syndromes including Werner's syndrome manifest elderly-looking symptoms earlier than usual. Studies of several progeroid syndromes have revealed a relationship to mutations in DNA metabolism enzymes such as helicases. Although WS is caused by the deficiency of RecQ3 helicase, the precise function of the enzyme remains uncharacterized. Interestingly, missense mutations of the gene coding for lamin A have been found in few cases of atypical WS patients (WS patient with no mutation in the WRN gene but exhibiting all the phenotypes associated with the syndrome).[52] Mutations in lamin A affects the organization of the nuclear lamina of cells. Different missense mutations have also been described for patients with the rare childhood progeroid syndrome, Hutchinson-Gilford progeria syndrome.[53] In the near future, the clarification of genotype/phenotype relationship in progeroid syndromes like WS will provide new insights into the genetic influence of human aging and pathophysiology of age-related diseases in general.

References

1. Epstein CJ, Martin GM, Schultz AL et al. Werner's syndrome: A review of its symptomatology, natural history, pathologic features, genetics and relationship to the natural aging process. Medicine 1966; 45:177-221.
2. Goto M, Tanimoto K, Horiuchi Y et al. Family analysis of Werner's syndrome: A survey of 42 Japanese families with a review of the literature. Clin Genet 1981; 19:8-15.
3. Goto M. Hierarchical deterioration of body systems in Werner's syndrome: Implications for normal ageing. Mech Age Dev 1997; 98:239-254.
4. Satoh M, Imai M, Sugimoto M et al. Prevalence of Werner's syndrome heterozygotes in Japan. Lancet 1999; 353:1766.
5. Martin GM. Genetic syndromes in man with potential relevance to the pathobiology of aging. Birth Defects 1978; 14:5-39.
6. Goto M, Rubenstein M, Weber J et al. Genetic linkage of Werner's syndrome to five markers on chromosome 8. Nature 1992; 355:735-738.
7. Yu C-E, Oshima J, Fu Y-H et al. Positional cloning of the Werner's syndrome gene. Science 1996; 272:258-262.
8. Werner O. On cataract in conjunction with scleroderma (doctoral dissertation, Kiel University). Schnidt and Klaunig, Kiel, 1904.
9. Oppenheimer BS, Kugel VH. Werner's syndrome: A heredofamilial disorder with scleroderma, bilateral juvenile cataract, precocious graying of hair and endocrine stigmatization. Trans Ass Amer Physicians 1934; 49:358-370.
10. Oppenheimer BS, Kugel VH. Werner's syndrome, report of the first necropsy and a findings in a new case. Am J Med Sci 1941; 202:629-642.
11. Agatson SA, Gartner S. Precocious cataracts and scleroderma (Rothmund's syndrome; Werner's syndrome). Arch Ophthalmol 1939; 21:492-496.
12. Thannhauser SJ. Werner's syndrome (progeria of the adult) and Rothmund's syndrome: Two types of closely related heredofamilial atrophic dermatosis with juvenile cataracts and endocrine features; a critical study with five new cases. Ann Intern Med 1945; 23:559-626.
13. Ishida R. A case of cataract associated with scleroderma. Jap J Ophthalmol 1917; 2:1025-1032.
14. Hayflick L, Moorhead PS. The serial cultivation of human diploid cell strains. Exp Cell Res 1961; 25:585-621.
15. Martin GM, Sprague CR, Epstein CJ. Replicative life-span of cultivated human cells. Lab Invest 1970; 23:86-92.
16. Martin GM, Gartler SM, Epstein CJ et al. Diminished lifespan of cultured cells in Werner's syndrome. Fed Proc 1965; 24:678.
17. Tokunaga M, Futami T, Wakamatsu E et al. Werner's syndrome as "hyaluronuria". Clin Chim Acta 1975; 62:89-92.
18. Goto M, Murata K. Urinary excretion of macromolecular acidic glycosaminoglycans in Werner's syndrome. Clin Chim Acta 1978; 85:101-106.
19. Maeda H, Fujita H, Sakura Y et al. A competitive enzyme-linked immunosorbent assay like method for the detection of urinary hyaluronan. Biosci Biotechnol Biochem 1999; 63:892-895.
20. Tanabe M, Goto M. Elevation of serum hyaluronan level in Werner's syndrome. Gerontol 2001; 47:77-81.
21. Hatamochi A. Dermatological features and collagen metabolism in Werner syndrome. In: Goto M, Miller RW, eds. From premature gray hair to helicase-Werner syndrome: Implications for aging and cancer. Tokyo: Karger, Japan Scientific Societies Press, 2001:51-59.
22. Salk D. Werner syndrome: A review of recent research with an analysis of connective tissue metabolism, growth control of cultured cells. In: Salk D, Fujiwara Y, Martin GM, eds. Werner's syndrome and human aging. New York and London: Plenum Press, 1985:215-218.
23. Gray MD, Shen JC, Kamath-Loeb AS et al. The Werner syndrome protein is a DNA helicase. Nat Genet 1997; 17:100-103.
24. Kitao S, Ohsugi I, Ichikawa K et al. Cloning of two new human helicase genes of RecQ family: Biological significance of multiple species in higher eukaryotes. Genomics 1998; 54:443-452.
25. Suzuki N, Shiratori M, Goto M et al. Werner syndrome helicase contains a 5'—>3' exonuclease activity that digests DNA and RNA strands in DNA/DNA and RNA/DNA duplexes dependent on unwinding. Nucleic Acids Res 1999; 27:2361-2368.
26. Drayna D. The search for the Werner syndrome gene. In: Goto M, Miller RW, eds. From premature gray hair to helicase-Werner syndrome: Implications for aging and cancer. Tokyo: Karger, Japan Scientific Societies Press, 2001:11-18.
27. Goto M, Horiuchi Y, Tanimoto K et al. Werner's syndrome: analysis of 15 cases with a review of the Japanese literature. J Am Geriatrics Soc 1978; 26:341-347.

28. Goto M, Imamura O, Kuromitsu J et al. Analysis of helicase gene mutations in Japanese Werner's syndrome patients. Hum Genet 1997; 99:191-193.
29. Matsumoto T, Imamura O, Yamabe Y et al. Mutation and haplotype analyses of the Werner's syndrome gene based on its genomic structure: genetic epidemiology in the Japanese population. Hum Genet 1997; 100:123-130.
30. Goto M, Yamabe Y, Shiratori M et al. Immunological diagnosis of Werner syndrome by down-regulated and truncated gene products. Human Genet 1999; 104:301-307.
31. Goto M, Horiuchi Y, Okumura K et al. Immunological abnormalities of aging: An analysis of T lymphocyte subpopulations of Werner's syndrome. J Clin Invest 1979; 64:695-699.
32. Goto M, Tanimoto K, Aotsuka S et al. Age-related changes in auto- and natural antibody in the Werner's syndrome. Am J Med 1982; 72:607-614.
33. Goto M, Tanimoto K, Horiuchi Y et al. Reduced natural killer cell activity of lymphocytes from patients with Werner's syndrome and recovery of its activity by purified human leukocyte interferon. Scand J Immunol 1982; 15:389-397.
34. Scammon RE. The measurement of the body in childhood. In: Harris JA, Jackson CM, Patterson DG et al, eds. The measurement of man. Minnesota: University of Minnesota Press, 1930:173-215.
35. Goto M, Kindynis P, Resnick D et al. Osteosclerosis of the phalanges in Werner's syndrome. Radiology 1989; 172:841-843.
36. Shiraki M, Aoki C, Goto M. Bone and calcium metabolism in Werner's syndrome. Endocrine J 1998; 45:505-512.
37. Kakehi T, Kuzuya H, Yoshimura Y et al. Binding and tyrosine kinase activities of the insulin receptor on Epstein-Barr virus transformed lymphocytes from patients with Werner's syndrome. J Gerontol 1988; 43:M40-45.
38. Takeuchi F, Kamatani N, Goto M et al. Gout-like arthritis in patients with Werner's syndrome. Jap J Rheumatol 1987; 1:214-220.
39. Goto M, Kato Y. Hypercoagulable state indicates an additional risk factor for atherosclerosis in Werner's syndrome. Thromb Haemost 1995; 73:576-578.
40. Goto M. Immunosenescent features of human segmental progeroid syndrome: Werner's syndrome. Aging Immunol Infect Dis 1992; 3:203-215.
41. Goto M, Nishioka K. Age- and sex-related changes of the lymphocyte subsets in healthy individuals: An anlysis by two-dimensional flow cytometry. J Gerontol 1989; 44:M51-56.
42. Goto M, Tanimoto K, Miyamoto T. Immunological aspects of the Werner's syndrome. In: Salk D, Fujiwara Y, Martin GM, eds. Werner's syndrome and human aging. New York and London: Plenum Press, 1985:263-284.
43. Sumi SM. Neuropathology of Werner syndrome. In: Salk D, Fujiwara Y, Martin GM, eds. Werner's syndrome and human aging. New York and London: Plenum Press, 1985:215-218.
44. Kakigi R, Endo C, Neshige R et al. Accelerated aging of the brain in Werner's syndrome. Neurology 1992; 42:922-924.
45. Goto M, Miller RW, Ishikawa Y et al. Excess of rare cancers in Werner's syndrome (adult progeria). Cancer Epidemiol Biomarker Prevention 1996; 5:239-246.
46. Fukuchi K, Tanaka K, Kumahara Y et al. K. Increased frequency of 6-thioguanine-resistant peripheral blood lymphocytes in Werner syndrome patients. Hum Genet 1990; 84:249-252.
47. Kyoizumi S, Kusunoki Y, Seyama T et al. In vivo somatic mutations in Werner's syndrome. Hum Genet 1998; 103:405-410.
48. Fraccaro M, Scappaticci S, Cerimele D. A population and cytogenetic study of the Werner syndrome in Sardinia. In: Salk D, Fujiwara Y, Martin GM, eds. Werner's syndrome and human aging. New York and London: Plenum Press, 1985:547-552.
49. Satoh M, Matsumoto T, Imai M et al. Prevalence of Werner syndrome gene mutations in the Japanese population: A genetic epidemiological study. In: Goto M, Miller RW, eds. From premature gray hair to helicase-Werner syndrome: Implications for aging and cancer. Tokyo: Karger, Japan Scientific Societies Press, 2001:19-25.
50. Matsumoto T, Tsuchihashi Z, Ito C et al. Genetic diagnosis of Werner's syndrome, a premature aging disease, by mutant allele specific amplification (MASA) and oligomer ligation assay (OLA). J Anti-Aging Med 1998; 1:131-140.
51. Goto M, Tanimoto K, Miyamoto T. Clinical, demographic, and genetic aspects of the Werner's syndrome in Japan. In: Salk D, Fujiwara Y, Martin GM, eds. Werner's syndrome and human aging. New York and London: Plenum Press, 1985:245-261.
52. Chen L, Lee L, Kudlow BA et al. LMNA mutations in atypical Werner's syndrome. Lancet 2003; 362:440-445.
53. Eriksson M, Brown WT, Gordon LB et al. Recurrent de novo point mutations in lamin A cause Hutchinson-Gilford progeria syndrome. Nature 2003; 423:293-298.

CHAPTER 2

Biochemical Roles of RecQ Helicases

Payam Mohaghegh and Ian D. Hickson

Abstract

The RecQ family of DNA helicases appears to influence DNA repair, replication and/or homologous recombination pathways. In humans, a defect in the RecQ family helicases encoded by the *BLM*, *WRN* and *RECQ4* genes gives rise to Bloom's, Werner's and Rothmund-Thomson syndromes, respectively. These disorders are associated with cancer predisposition and/or premature aging. In Bloom's syndrome, affected individuals are predisposed to many types of cancer at an early age. Werner's syndrome is a premature aging disorder with a complex phenotype, which includes many age-related disorders that develop from puberty, including graying and thinning of the hair, bilateral cataract formation, type II diabetes mellitus, osteoporosis and atherosclerosis. The phenotype of Rothmund-Thomson syndrome patients also consists of some features associated with premature aging, as well as predisposition to certain cancers. Here, we discuss the molecular and possible biochemical basis of these RecQ helicase-deficient disorders.

RecQ Helicases: A Family of Molecular Motor Proteins

A common feature of many processes in DNA metabolism is a requirement for the Watson and Crick strands of a DNA duplex to be separated through the action of a helicase. DNA helicases are ubiquitous in nature and utilize the energy derived from ATP hydrolysis to perform numerous roles in genetic recombination, gene transcription, DNA replication and DNA repair.[1] Recent data indicate that germ-line mutations in human genes encoding members of one particular family of DNA helicases, the RecQ family, give rise to hereditary disorders of man associated with features of premature aging and/or cancer predisposition.[2] The RecQ family is highly conserved in evolution from prokaryotes through to mammals (Fig. 1), and is named after the RecQ protein of *Escherichia coli*. RecQ family members share a highly conserved domain comprising approximately 450 amino acids that includes seven sequence motifs found in many classes of DNA and RNA helicases. Amongst these motifs is an ATP binding sequence (the so-called Walker A-box; a characteristic of NTP-binding enzymes) and a DEAH box. The existence of these motifs alone does not, however, provide conclusive evidence that the particular protein will act as a helicase. Certain proteins containing these motifs do not separate the complementary strands of DNA, but instead translocate along nucleic acid molecules. Outside of the helicase domain there is only limited sequence similarity amongst RecQ family members. However, certain other conserved sequence features are evident in RecQ family members (Fig. 1). Where studied, RecQ family members have been shown to be *bona fide* DNA helicases that translocate in the 3'-5' direction relative to the strand to which they are

---

Molecular Mechanisms of Werner's Syndrome, edited by Michel Lebel. ©2004 Eurekah.com and Kluwer Academic / Plenum Publishers.

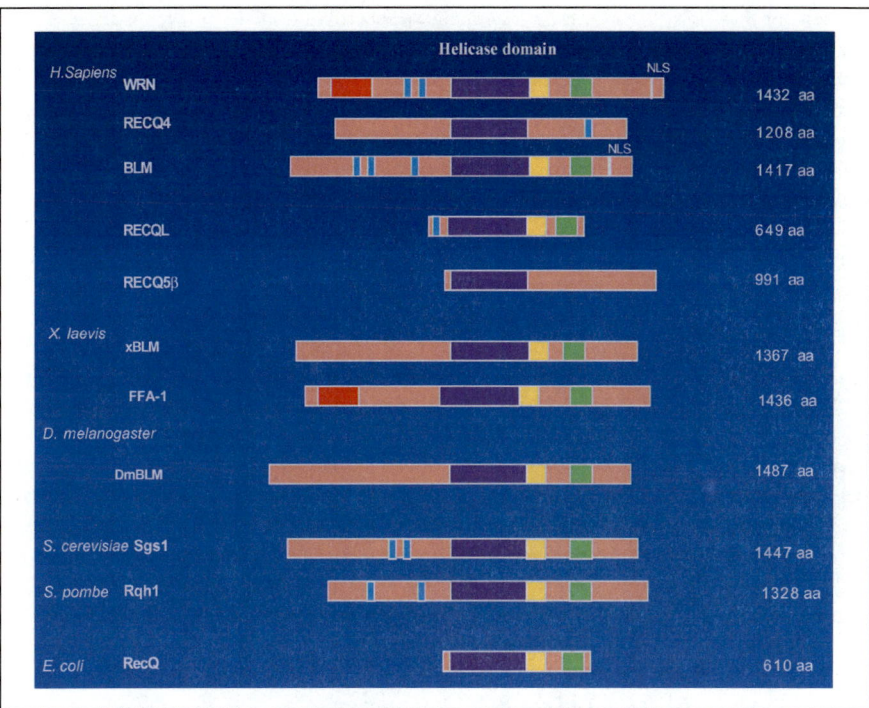

Figure 1. Schematic representation of the RecQ family of DNA helicases. Members of this family have been identified in bacteria (*E.coli*: RecQ), yeasts (*Saccharomyces cerevisiae*: Sgs1; *Schizosaccharomyces pombe*: Rqh1) fruit fly (*Drosophila melanogaster:* DmBLM), frog (*Xenopus laevis*: the BLM orthologue, xBLM, and the WRN orthologue, FFA-1), and mammals - only human is shown (*Homo sapiens*: BLM, WRN, RECQ4, RECQL, RECQ5). The central conserved helicase domain is shown as a purple box. The exonuclease domain of WRN and FFA-1 (which is not found in other RecQ helicases) is shown as a red box. The yellow boxes denote a region of conservation C-terminal to the helicase domain. This region is apparently unique to RecQ helicases. The green boxes show the so-called HRDC domain, which is also found in certain enzymes that degrade RNA, and the blue boxes show highly acidic regions. The gray boxes denote domains with little or no sequence similarity between family members. The nuclear localization signal sequences (NLS) in BLM and WRN are shown in black and marked above. Three variants of the RECQ5 protein may exist, only the β isoform is indicated. The size of each protein (in amino acid residues) is shown on the right.

bound.[3] These enzymes, like all helicases, convert the chemical energy derived from NTP hydrolysis into mechanical energy in the form of translocation relative to the DNA backbone, leading to separation of the complementary strands. Hence, helicases can be considered to be 'molecular motor proteins'.[1] WRN is unique amongst the RecQ helicases in also being a 3'-5' exonuclease, an activity that is dependent upon a domain in the N-terminal region of the protein (Fig. 1).[4] WRN has, therefore, a bipartite structure; the WRN helicase domain being functionally distinct and separable from the exonuclease domain. However, these two domains seem to work in concert in the holoenzyme.[5]

RecQ Helicase Diseases

Defects in RecQ helicases in humans lead to genomic instability disorders associated with cancer predisposition and/or premature aging.[2-4] These disorders are Bloom's Syndrome (BS), Werner's syndrome (WS), and Rothmund-Thomson syndrome (RTS), which are caused by mutations in the *BLM, WRN* and *RECQ4* genes, respectively (Fig. 1). In each case, cells derived from affected individuals show chromosomal instability that manifests even in cells grown in the absence of DNA damaging agents. Based on analysis of the mutations found in BS, WS and RTS patients, it is clear that loss of function mutations in *BLM, WRN* or *RECQ4* are compatible with embryonic development, which poses the question of possible functional redundancy between different human RecQ helicases. Despite the apparent structural and biochemical similarities between the BLM, WRN and RECQ4 proteins,[2,6] the phenotypes of BS, WS and RTS are obviously different[2,4,7,8] suggesting that each disease pathway is functionally distinct, at least to some extent. However, it is also clear that these disorders share certain phenotypic characteristics. For example, although premature aging is the defining feature of WS, BS and RTS patients also show a limited selection of aging phenotypes, such as the premature development of cataracts or Type II diabetes. Moreover, the premature development of cancers in BS, WS and RTS can be viewed as one symptom of premature aging.

Recently, another clear link between DNA helicase deficiency and premature aging has been demonstrated. The rare disorder, xeroderma pigmentosum, is caused by mutations in any one of seven genes (designated *XPA-G*). Some mutations in the *XPB* or *XPD* genes also cause Cockayne's syndrome (CS) and trichothiodystrophy (TTD).[9-10] It was shown that a mutation in *XPD*, which encodes a DNA helicase that is involved in DNA repair and transcription through being a component of the transcription complex TFIIH, leads to accumulation of DNA damage and symptoms of premature aging. Double mutants of *XPD* and *XPA* (also involved in DNA repair, but not apparently in transcription) showed an even faster rate of aging.[11]

Mouse Models of RecQ Helicase Deficiency

Mouse models of RecQ helicase deficiency have been generated. Cells from BLM-/- mice show inherent genomic instability.[12,13] Although in only one study were viable animals recovered, these mice are mildly cancer-prone.[13] Moreover, ES cells lacking BLM show an increased frequency of gene targeting, indicative of an increase in homologous recombination (HR). This is consistent with the fact that HR mediates sister-chromatid exchanges (SCEs) in vertebrates, and that elevated SCEs are seen in BS cells. An increased level of recombination between homologous chromosomes in BS cells may lead to an elevation in the rate of somatic loss-of-heterozygosity, which in turn would lead to inactivation of tumor suppressor genes and hence to cancer predisposition in BS.[13] *WRN*–defective mice have also been generated, but do not display obvious signs of premature aging. These mice have been crossed with p21 or p53 null mice to generate double mutants.[14] At least in the first few generations, these double mutant mice also do not display premature aging. The p53-*WRN* double mutant mice do, however, show an accelerated rate of tumorigenesis associated with the development of a variety of tumors not commonly seen in either of the single mutant mice.[14] In a separate study,[15] transgenic mice expressing a mutant form of *WRN* encoding a defective helicase were shown to exhibit none of the premature aging phenotypes characteristic of WS. However, primary cell cultures from these mice display two of the well established features of WS cells; reduced replicative potential and hypersensitivity to camptothecin. It is possible that the lack of obvious signs of premature aging in these mouse models reflects the fact that mouse cells have very long telomeres, the specialized structures that cap the ends of linear eukaryotic chromosomes. One of the factors that may influence the aging process in mammals is the shortening of telomeres (discussed in more detail below). Human chromosomes have much shorter telomeres than mouse chromosomes.

Possible Role for WRN in DNA Replication/Repair

There is increasing evidence to suggest that WRN functions in some capacity in DNA replication and in cellular responses to DNA damage. Amongst the evidence supporting this proposal are the following observations.[2,4,7,16-23] (1) WRN translocates to presumptive sites of replication/repair (defined by nuclear 'foci' containing certain proteins such as replication protein A (RPA)) when DNA replication is blocked by the ribonucleotide reductase inhibitor hydroxyurea. (2) The *Xenopus laevis* WRN homologue, which is known as focus-forming activity-1, is essential for the formation of replication foci containing replication 'factories'. (3) WRN copurifies with a multi-enzyme DNA replication complex and interacts physically with replication proteins, including proliferating cell nuclear antigen (PCNA), FEN1 and topoisomerase I. (4) The heterotrimeric, single-stranded DNA binding protein, RPA, interacts with WRN, and its presence increases WRN helicase activity. (5) WS cells show a protracted S-phase. (6) WRN interacts functionally with DNA polymerase δ, which is required for DNA replication. (7) WRN can stimulate the nuclease activity of FEN-1, which also plays roles in DNA replication (Okazaki fragment processing) and DNA repair (Fig. 2).[23]

WRN forms a complex with the p53 tumor suppressor protein and p53-dependent apoptosis is defective in WS cells.[24-26] In normal cells, the p53-WRN complex may recognize abnormal DNA structures, such as stalled replication forks, leading either to the coordination of cell cycle and DNA repair events, or to the induction of apoptosis. WRN could be involved directly in restoration of DNA replication by displacing (perhaps abnormal?) Okazaki fragments on the lagging strand and degrading the displaced DNA. Recently, it has been shown that the WRN exonuclease catalyses structure-dependent degradation of DNA,[27] suggesting that WRN resolves abnormal DNA structures via both its helicase and its exonuclease activities. Moreover, WRN interacts with several proteins required for lagging strand synthesis, including PCNA and FEN1. A recent paper[28] supports a role for the WRN exonuclease activity during repair processes. It was shown the Ku70 and Ku86 DNA end-binding complex directly interacts with WRN, and stimulates its 3'-5' exonuclease activity.

A second possible role for WRN in replication fork repair would be after removal of the damaged DNA strand at blocked forks. For example, WRN may assist the recombination machinery in the reinitiation of a productive replication fork from a recombination intermediate such as a D-loop (discussed in more detail in ref. 29) Indeed, WRN relocates into nuclear foci in response to DNA damaging agents and colocalizes with RPA and RAD51.[30,31]

The interaction of WRN with DNA polymerase δ provides a direct biochemical link between WRN and DNA synthesis. It was shown that WRN increases the rate of nucleotide incorporation by DNA polymerase δ in the absence of PCNA; however, WRN has no significant stimulatory effect on the DNA polymerase δ holoenzyme (polymerase δ PCNA complex).[22] Therefore, Kamath-Loeb et al[22] suggested that WRN is unlikely to function in normal, processive DNA synthesis, which requires the polymerase δ PCNA complex. Rather, they speculated that WRN may function in a replication restart pathway at sites where damaged DNA/unusual DNA secondary structures have blocked DNA replication and where the DNA replication machinery has detached from the DNA. It has also been shown that WRN is able to recruit DNA polymerase δ to the nucleolus, suggesting that WRN could be involved in regulating the initiation and progression of DNA replication by recruiting polymerase δ to particular sites of DNA synthesis (Fig. 2).[32] Interestingly, PCNA binds the exonuclease domain of WRN and appears to drive the oligomerization of the exonuclease region of WRN, and to stimulate the cleavage of the 5'protruding strand at a single strand-double strand DNA junction.[33] PCNA is normally present as a trimer, and the exonuclease domain of WRN is also present as a trimer and as a hexamer. PCNA trimers and WRN trimers may form a combined heteromeric hexamer in presence of certain DNA substrates.

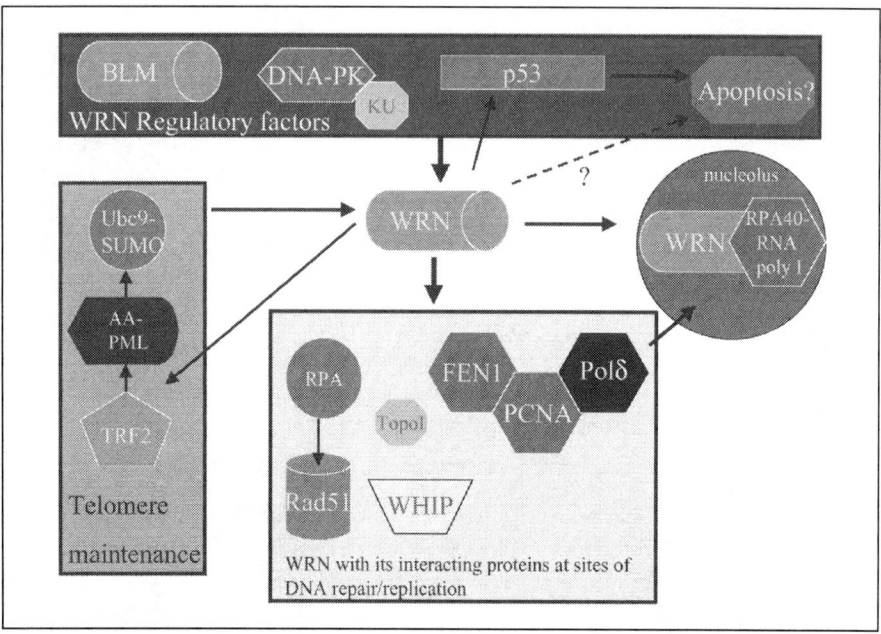

Figure 2. Nuclear interactions involving WRN. The box with a dark background represents WRN regulatory factors. BLM inhibits WRN exonuclease function. WRN and p53 physically interact via their C-terminal domains, and the processing of Holliday junctions by WRN helicase is regulated by p53.[53] Moreover, p53-mediated apoptosis is defective in WS fibroblasts; however, the exact pathway is not clear and hence is marked '?'. WRN appears to be a coactivator of p53-meditated transcription. DNA-PK interacts with and phosphorylates WRN. Ku stimulates WRN exonuclease activity. WRN localisation to AA-PML bodies might also be modulated by p53. In turn, WRN might influence targeting of TRF1/TRF2 to AA-PML bodies and modulate their involvement in telomere maintenance. These events probably depend on SUMO modification; this idea is supported by the finding that WRN and Ubc9 interact directly. The pathway involving p53-WRN/TRF1/2-AAPML appears to be crucial in the maintenance of telomere ends and could have implications in cellular aging. WRN localizes to the nucleolus. WRN interaction with the RPA40 unit of RNA polymerase is needed for rRNA transcription in the nucleolus. WRN interacts with number of proteins that are involved DNA repair and replication: FEN1-PCNA-Pol δ, RPA-Rad51, TOPOI and the novel gene WHIP.

The proposed role of WRN in replication was further emphasized by the recent identification of WHIP (Werner Helicase Interacting Protein) as a WRN-binding factor. WHIP shows homology to the replication factor C family of proteins, and is conserved from *E. coli* to human cells.[34] Budding yeasts (*Saccharomyces cerevisiae*) deficient in WHIP (yWHIP) show a reduced lifespan, an effect that is additive to the previously described premature aging phenotype of *sgs1* mutants.

Roles for WRN and other RecQ helicases outside of the process of DNA replication and repair also have been proposed. A recently described novel role for WRN was suggested by Shiratori et al[35] WRN appears to accelerate the transcription of ribosomal RNA as a component of an RNA polymerase I-associated complex. WRN has a nucleolar localization signal, and is apparently involved in intranuclear trafficking between the nucleolus and the nucleoplasm (Fig. 2). How this role explains the genomic instability evident in WS cells remains to be explored in detail.

Regulation of WRN Activity

Although the mechanism by which regulation of WRN activity is effected remains to be worked out in detail, there have been some reports detailing potential pathways for WRN regulation. It was shown recently[36,37] that WRN interacts with and is a target for the DNA-dependent protein kinase (DNA-PK). This phosphorylation may require the presence of Ku protein, and Ku interacts with WRN and stimulates its exonuclease activity. There are, however, somewhat conflicting reports of the effect that DNA-PK has on WRN function. Yannone et al[37] found that DNA-PK catalytic subunit can inhibit both the exonuclease and helicase activities of WRN; an effect reversed by binding of Ku to the complex. In contrast, Karmakar et al[36] showed that the DNA-PK catalytic subunit alone does not affect WRN exonuclease activity without the presence of Ku. Nevertheless, these authors did show that DNA-PK mediated phosphorylation of a previously dephosphorylated WRN preparation did inhibit catalytic activity, consistent with the results of Yannone et al.[37]

Another potential way of regulating and coordinating the many functions of WRN is by SUMO (small ubiquitin-related modifier) modification (Fig. 2). Conjugation of SUMO to target proteins appears to regulate subcellular localization in some cases, the half-life of the protein, as well as possibly modulating interactions with other proteins. WRN has been shown to be covalently modified by SUMO-1 via the conjugating enzyme Ubc9.[38] Ubc9 has been shown to play a role in the degradation of certain proteins, including S- and M-phase cyclins.[39] Moreover, SUMO modification is involved in regulating p53 function, through interactions involving the Mdm2 protein and changes in the half-life of p53.[40] Potentially, SUMO modification of p53 and WRN could play a critical role in orchestrating the cross talk between these proteins and hence in regulating pathways for the maintenance of genome integrity.

WRN and BLM protein also physically interact and colocalize with each other, and BLM is able to inhibit the exonuclease activity of WRN.[41] This report supported the proposition that RecQ family members may be able to influence each other in ways that could mask the true phenotype of mutation of the *WRN* or *BLM* gene. Hence, it would be interesting to examine the effect of loss of both BLM and WRN in mice.

A Role for RecQ Helicases in Telomere Maintenance

Several lines of evidence suggest a link between telomere erosion and the induction of cellular senescence in eukaryotes. Telomeres are the repetitive DNA structures at the tips of every chromosome.[42] Their length and integrity are maintained through the action of numerous proteins, of which the specialized reverse transcriptase, telomerase, plays a vital role. In proliferating cells, the failure to replicate to the very end of each chromosome (the so-called 'end-replication problem') would lead to progressive telomere erosion were it not for the ability of telomerase to extend telomere ends utilizing an intrinsic RNA component as a template. Telomerase also plays another, rather poorly defined, role in telomere 'capping', which seems to prevent telomeric ends from being seen by the cells as DNA damage (i.e., as DNA breaks) and hence invoking cell cycle checkpoint responses. In the absence of telomerase (as in human somatic cells, or in mutant cells lacking telomerase) telomeres erode gradually. This eventually leads to cellular senescence unless telomerase can be reactivated by some means. However, in cells lacking telomerase, 'survivors' emerge from this state at some still ill-defined frequency, and these are able to proliferate in the absence of telomerase using a recombination-mediated process of telomere maintenance, termed ALT (for alternative lengthening of telomeres) in mammals. ALT cells tend to have long telomeres of variable length.[42]

It is possible that WRN plays a role in the maintenance of telomere integrity. Fibroblasts derived from WS patients show an accelerated rate of replicative senescence, but the over-expression of telomerase in these cells halts this premature senescence.[43] In most human tumors, the gene encoding telomerase is reactivated and hence maintains telomere integrity during continued proliferation. However, in ~10% of tumors, telomerase is not reactivated and telomeres are maintained by the ALT process, which may require WRN. The best evidence linking RecQ helicases with this process is the fact that the WRN homologue in *Saccharomyces cerevisiae*, *SGS1*, is involved in telomere maintenance, at least in strains lacking telomerase.[44-46] Deletion of *EST2*, which encodes the catalytic subunit of telomerase in yeast, leads to premature replicative senescence, and a progressive shortening of telomeres. The rare survivors emerging from such a population show either amplification of subtelomeric DNA elements, or long and variable telomeric TG_{1-3} repeat sequences reminiscent of those seen in human ALT cells. Deletion of *SGS1* in *est2* strains leads to an accelerated rate of senescence. Moreover, *est2 sgs1* strains do not generate 'ALT-like' survivors, indicating that Sgs1p is required for telomere lengthening in the absence of telomerase. The role of Sgs1p in telomeric function is apparently conserved in evolution, since mouse *WRN* can partially substitute for *SGS1*.[45] This suggests that WRN may play a similar role in humans to that played by Sgs1p in yeast. The recombination/repair genes, *RAD51* and *RAD52*, are essential for cell survival in *sgs1 est2* double mutant strains, indicating that Sgs1p participates in the poorly characterized, Rad50p-dependent recombination pathway for telomere maintenance.[46] In order to gain insight into the role of RecQ helicases in telomere biology, a detailed study of their biochemical properties is essential.

Biochemical Properties of RecQ Helicases

BLM and WRN are clearly atypical helicases and are highly DNA structure specific in their action. For example, neither BLM nor WRN is capable of unwinding duplex DNA from a blunt-ended terminus, or from an internal nick. However, both enzymes efficiently unwind the same blunt-ended duplex containing a centrally located 12 nt single-stranded 'bubble'. Moreover, WRN and BLM can disrupt a synthetic X-structure (a model for the Holliday junction recombination intermediate) in which each 'arm' of the 4-way junction is blunt-ended, indicating that these enzymes specifically recognize the cross-over structure.[47] Consistent with this, both BLM and WRN can promote branch migration of Holliday junctions.[31,47] BLM and WRN also both unwind a variety of different forms of G-quadruplex DNA[6,48,49] a stable secondary structure that can form at guanine-rich sequences present at several genomic loci.[50,51] Of particular relevance to this article is the proposal that G-quaduplex DNA can form readily within telomeric repeat DNA.[42,50] It is possible, therefore, that one role for RecQ helicases in telomere maintenance is to disrupt G-quaduplexes to assist in the replication and/or repair of telomeric DNA. WRN and BLM are also able to unwind triple helix DNA,[52] another unusual DNA structure that might be present under certain circumstances at telomere ends; this unwinding requires a 3' tail attached to the third strand. With the exception that WRN is an exonuclease, the BLM and WRN helicases appear to be very similar in their substrate specificity.[6]

Summary

In summary, RecQ helicases play a vital role in the maintenance of genome integrity. Deficiency in RecQ helicases in man leads to a variety of features of premature aging. However, although there are some specific biochemical features of RecQ helicases that could explain a role in aging, such as their association with DNA repair proteins and an ability to disrupt G-quadruplexes that form within telomeric DNA, we are still lacking definitive evidence of a causative role for these enzymes in the prevention of premature replicative senescence.

Acknowledgements

We thank Dr L. Wu for helpful comments, Mrs J. Pepper for preparation of the manuscipt, and Cancer Research UK for financial support.

References

1. Soultanas P, Wigley DB. Unwinding the 'Gordian knot' of helicase action. Trends Biochem Sci 2001; 26:47-54.
2. Mohaghegh P, Hickson ID. DNA helicase deficiencies associated with cancer predisposition and premature ageing disorders. Hum Mol Genet 2001; 10:741-746.
3. Karow JK, Wu L, Hickson ID. RecQ family helicases: Roles in cancer and aging. Curr Opin Genet Dev 2000; 10:32-38.
4. Shen JC, Loeb LA. The Werner syndrome gene: The molecular basis of RecQ helicase- deficiency diseases. Trends Genet 2000; 16:213-220.
5. Machwe A, Xiao L, Theodore S et al. DNase I footprinting and enhanced exonuclease function of the bipartite Werner syndrome protein (WRN) bound to partially melted duplex DNA. J Biol Chem 2002; 277:4492-4504.
6. Mohaghegh P, Karow JK, Brosh RM et al. The Bloom's and Werner's syndrome proteins are DNA structurespecific helicases. Nucleic Acids Res 2001; 29:2843-2849.
7. Shen J, Loeb LA. Unwinding the molecular basis of the Werner syndrome. Mech Ageing Dev 2001; 122:921-944.
8. Levitt NC, Hickson ID. Caretaker tumour suppressor genes that defend genome integrity. Trends Mol Med 2002; 8:179-186.
9. Vermeulen W, Scott RJ, Rodgers S et al. Clinical heterogeneity within xeroderma pigmentosum associated with mutations in the DNA repair and transcription gene ERCC3. Am J Hum Genet 1994; 54:191-200.
10. Lehmann AR. The xeroderma pigmentosum group D (XPD) gene: One gene, two functions, three diseases. Genes Dev 2001; 15:15-23.
11. de Boer J, Andressoo JO, de Wit J et al. Premature aging in mice deficient in DNA repair and transcription. Science 2002; 296:1276-1279.
12. Chester N, Kuo F, Kozak C et al. Stage-specific apoptosis, developmental delay, and embryonic lethality in mice homozygous for a targeted disruption in the murine Bloom's syndrome gene. Genes Dev 1998; 12:3382-3393.
13. Luo G, Santoro IM, McDaniel LD et al. Cancer predisposition caused by elevated mitotic recombination in Bloom mice. Nat Genet 2000; 26:424-429.
14. Lebel M, Cardiff RD, Leder P. Tumorigenic effect of nonfunctional p53 or p21 in mice mutant in the Werner syndrome helicase. Cancer Res 2001; 61:1816-1819.
15. Wang L, Ogburn CE, Ware CB et al. Cellular Werner phenotypes in mice expressing a putative dominant- negative human WRN gene. Genetics 2000; 154:357-362.
16. Yan H, Chen CY, Kobayashi R et al. Replication focus-forming activity 1 and the Werner syndrome gene product. Nat Genet 1998; 19:375-378.
17. Lebel M, Spillare EA, Harris CC et al. The Werner syndrome gene product copurifies with the DNA replication complex and interacts with PCNA and topoisomerase I. J Biol Chem 1999; 274:37795-37799.
18. Brosh Jr RM, Orren DK et al. Functional and physical interaction between WRN helicase and human replication protein A. J Biol Chem 1999; 274:18341-18350.
19. Poot M, Hoehn H, Runger TM et al. Impaired S-phase transit of Werner syndrome cells expressed in lymphoblastoid cell lines. Exp Cell Res 1992; 202:267-273.
20. Ogburn CE, Oshima J, Poot M et al. An apoptosis-inducing genotoxin differentiates heterozygotic carriers for Werner helicase mutations from wild-type and homozygous mutants. Hum Genet 1997; 101:121-125.
21. Brosh Jr RM, Karow JK et al. Potent inhibition of werner and bloom helicases by DNA minor groove binding drugs. Nucleic Acids Res 2000; 28:2420-2430.
22. Kamath-Loeb AS, Johansson E, Burgers PM et al. Functional interaction between the Werner Syndrome protein and DNA polymerase delta. Proc Natl Acad Sci USA 2000; 97:4603-4608.
23. Brosh Jr RM, von Kobbe C, Sommers JA et al. Werner syndrome protein interacts with human flap endonuclease 1 and stimulates its cleavage activity. Embo J 2001; 20:5791-5801.
24. Spillare EA, Robles AI, Wang XW et al. p53-mediated apoptosis is attenuated in Werner syndrome cells. Genes Dev 1999; 13:1355-1360.

25. Blander G, Kipnis J, Leal JF et al. Physical and functional interaction between p53 and the Werner's syndrome protein. J Biol Chem 1999; 274:29463-29469.
26. Robles AI, Harris CC. p53-mediated apoptosis and genomic instability diseases. Acta Oncol 2001; 40:696-701.
27. Shen JC, Loeb LA. Werner syndrome exonuclease catalyzes structuredependent degradation of DNA. Nucleic Acids Res 2000; 28:3260-3268.
28. Cooper MP, Machwe A, Orren DK et al. Ku complex interacts with and stimulates the Werner protein. Genes Dev 2000; 14:907-912.
29. Oakley TJ, Hickson ID. Defending genome integrity during S-phase: putative roles for RecQ helicases and topoisomerase III. DNA Repair 2002; 19:175-207.
30. Sakamoto S, Nishikawa K, Heo SJ et al. Werner helicase relocates into nuclear foci in response to DNA damaging agents and colocalizes with RPA and Rad51. Genes Cells 2001; 6:421-430.
31. Constantinou A, Tarsounas M, Karow JK et al. Werner's syndrome protein (WRN) migrates Holliday junctions and co localizes with RPA upon replication arrest. EMBO Rep 2000; 1:80-84.
32. Szekely AM, Chen YH, Zhang C et al. Werner protein recruits DNA polymerase delta to the nucleolus. Proc Natl Acad Sci USA 2000; 97:11365-11370.
33. Xue Y, Ratcliff GC, Wang H et al. A minimal exonuclease domain of WRN forms a hexamer on DNA and possesses both 3'- 5' exonuclease and 5'-protruding strand endonuclease activities. Biochemistry 2002; 41:2901-2912.
34. Kawabe Y, Branzei D, Hayashi T et al. A novel protein interacts with the Werner's syndrome gene product physically and functionally. J Biol Chem 2001; 276:20364-20369.
35. Shiratori M, Suzuki T, Itoh C et al. WRN helicase accelerates the transcription of ribosomal RNA as a component of an RNA polymerase I-associated complex. Oncogene 2002; 21:2447-2454.
36. Karmakar P, Piotrowski J, Brosh RM et al. Werner protein is a target of DNA-dependent protein kinase in vivo and in vitro, and its catalytic activities are regulated by phosphorylation. J Biol Chem 2002; 277:18291-18302.
37. Yannone SM, Roy S, Chan DW et al. Werner syndrome protein is regulated and phosphorylated by DNA- dependent protein kinase. J Biol Chem 2001; 276:38242-38248.
38. Kawabe Y, Seki M, Seki T et al. Covalent modification of the Werner's syndrome gene product with the ubiquitin-related protein, SUMO-1. J Biol Chem 2000; 275:20963-20966.
39. Seufert W, Futcher B, Jentsch S. Role of a ubiquitin-conjugating enzyme in degradation of S- and M-phase cyclins. Nature 1995; 373:78-81.
40. Hochstrasser M. Biochemistry. All in the ubiquitin family. Science 2000; 289:563-564.
41. Von Kobbe C, Karmakar P, Dawut L et al. Colocalization, Physical, and Functional Interaction between Werner and Bloom Syndrome Proteins. J Biol Chem 2002; 277:22035-22044.
42. McEachern MJ, Krauskopf A, Blackburn EH. Telomeres and their control. Annu Rev Genet 2000; 34:331-358.
43. Wyllie FS, Jones CJ, Skinner JW et al. Telomerase prevents the accelerated cell ageing of Werner syndrome fibroblasts. Nat Genet 2000; 24:16-17.
44. Johnson FB, Marciniak RA, McVey M et al. The Saccharomyces cerevisiae WRN homolog Sgs1p participates in telomere maintenance in cells lacking telomerase. Embo J 2001; 20:905-913.
45. Cohen H, Sinclair DA. Recombination-mediated lengthening of terminal telomeric repeats requires the Sgs1 DNA helicase. Proc Natl Acad Sci USA 2001; 98:3174-3179.
46. Huang P, Pryde FE, Lester D et al. SGS1 is required for telomere elongation in the absence of telomerase. Curr Biol 2001; 11:125-129.
47. Karow JK, Constantinou A, Li JL et al. The Bloom's syndrome gene product promotes branch migration of holliday junctions. Proc Natl Acad Sci USA 2000; 97:6504-6508.
48. Sun H, Karow JK, Hickson ID et al. The Bloom's syndrome helicase unwinds G4 DNA. J Biol Chem 1998; 273:27587-27592.
49. Fry M, Loeb LA. Human werner syndrome DNA helicase unwinds tetrahelical structures of the fragile X syndrome repeat sequence d(CGG)n. J Biol Chem 1999; 274:12797-12802.
50. Williamson JR. G-quartet structures in telomeric DNA. Annu Rev Biophys Biomol Struct 1994; 23:703-730.
51. Sen D, Gilbert W. A sodium-potassium switch in the formation of four-stranded G4-DNA. Nature 1990; 344:410-414.

52. Brosh Jr RM, Majumdar A, Desai S et al. Unwinding of a DNA triple helix by the Werner and Bloom syndrome helicases. J Biol Chem 2001; 276:3024-3030.
53. Yang Q, Zhang R, Wang XW et al. The processing of Holliday junctions by BLM and WRN helicases is regulated by p53. J Biol Chem 2002; 277:31980-31987.

CHAPTER 3

Biochemical Characterization of the Werner Syndrome DNA Helicase-Exonuclease

Michael Fry

Abstract

The positional cloning in 1996 of *WRN*, the human gene defective in Werner syndrome (WS), launched an explosive experimental activity that culminated in the expression, purification and comprehensive characterization of its product protein—the WRN DNA helicase-exonuclease. Being a member of the RecQ family of DNA helicases, WRN possesses a 3'→5' helicase activity that resides in a 7-motif domain at the center of the 1432 amino acid polypeptide. In addition, WRN is distinguished from all the other RecQ-like helicases in that it also has a 3'→5' exonuclease activity sited in a 3-motif N-terminal domain in the WRN protein. A distinctive feature of the helicase activity of WRN is its capacity to preferentially unwind alternate DNA structures that include tetra- and triple-helical DNA, duplex DNA containing a single-strand bubble, and 4-way X junction DNA. Correspondingly, the WRN 3'→5' exonuclease is also distinguished by its proclivity to preferentially digest non-canonical DNA structures such as bubble-containing duplex DNA, extra-helical single-strand loops and 3- and 4-way X junctions. The penchant of both the helicase and exonuclease for alternative DNA structures might reflect their roles in resolving aberrant DNA formations that form during the metabolism of cellular DNA. Some data indicate that the WRN helicase and nuclease act in coordination. However, it is not clear yet how the two activities operate in concert despite their opposite directions of advancement along the processed DNA molecule. Data described elsewhere in this volume indicate that the association of WRN protein with various DNA metabolism proteins modulates the innate properties of its helicase and exonuclease activities. Conceivably, therefore, a better understanding of the modes of operation and in vivo roles of the WRN protein are to be gained by studying the properties of its helicase and exonuclease in reconstituted multiprotein complexes.

The Human *WRN* Gene and Its Homologs

The *WRN* Gene

The human *WRN* locus was initially localized to 8p12 by linkage analysis[1] and its position was subsequently refined within an 8.3 centimorgan interval by meiotic and homozygosity mapping.[2-5] These data and the identification of markers that were found to be in linkage disequilibrium with WS in Japanese patients[6,7] formed the basis for the positional cloning in 1996 of *WRN* by Schellenberg and colleagues.[8] The identification of multiple mutations in *WRN* of WS patients established *WRN* as the gene responsible for WS.[8,9] The currently docu-

Molecular Mechanisms of Werner's Syndrome, edited by Michel Lebel. ©2004 Eurekah.com and Kluwer Academic / Plenum Publishers.

Biochemical Characterization of the Werner Syndrome DNA Helicase-Exonuclease

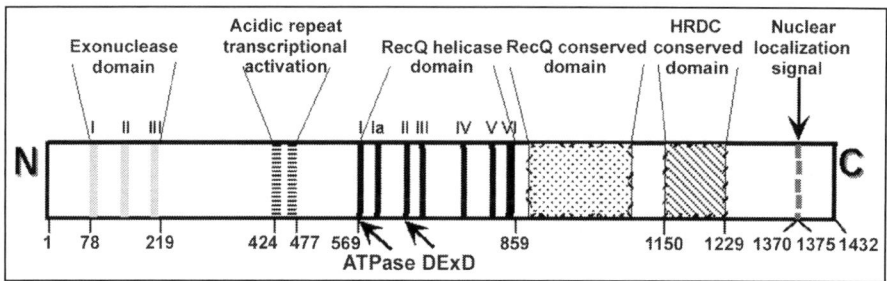

Figure 1. The WRN helicase-exonuclease. The schematically represented domains identified in the 1432-amino acid protein are: A centrally located (residues 569-859) seven-motif RecQ helicase domain; an N-terminal (residues 78-219) three-motif exonuclease domain; a C-terminal (residues 1370-1375) nuclear localization signal (NLS); a transcription activating 27-amino acid direct repeat located between the helicase and exonuclease domains (residues 424-477). The two additional elements situated between the helicase domain and the NLS are the RecQ conserved domain and the helicase and RNase D C-terminal domain (HRDC).

mented domains in the WRN protein are schematically depicted in Figure 1 and are briefly described in this section.

WRN cDNA encodes a protein of 1432 amino acids that is highly similar to DNA helicases of the RecQ family. A centrally located helicase region (amino acids 569-869) consists of seven motifs shared by every known RecQ-type helicase. Motif I contains a nucleotide binding ATPase sequence and domain II includes a DExH tract (Asp-Glu-any amino acid-His), marking WRN as a member of the RecQ family of DNA helicases.[8] Subsequent to the realization that WRN encodes a DNA helicase, its gene product the WRN[1] protein, was expressed in *Spodoptera frugiperda* insect cells and purified by affinity column chromatography. The isolated protein was shown to contain both ATPase and DNA unwinding activities characteristic of DNA helicase.[10,11]

A nucleotide tract located upstream to the helicase region encodes an acidic stretch of 98 amino acids. This domain contains a perfectly duplicated 27-residue repeat (amino acids 424-477). Subsequently published results suggested a role for this repeat in the activation of transcription by WRN (refs. 12, 13, vide infra).

Employment of iterative strategy combining several sensitive computer methods[14,15] and Hidden Markov Model analysis[16] of the protein amino acids sequence, revealed sequence homology between an N-terminal three-motif region of WRN (residues 78 to 219) and several 3'→5' nucleases. Subsequent to this analysis, results of biochemical and genetic experiments provided a direct demonstration that this domain indeed possesses 3'→5' exonuclease activity (refs. 17-19, see below).

A C-terminal 63-amino acids nuclear localization signal sequence (NLS) in WRN was detected by following the migration to the nucleoplasm of protein products of N-terminal truncated mutants of WRN fused to an EGF-encoding reporter sequence.[20] The basic residues Lys[1371]-Arg[1372]-Arg[1373] within the NLS region were identified as essential for the transportation of the WRN protein to the cell nucleus.[20,21]

Sequence homology analysis revealed two additional domains in the C-terminal section of WRN. Closer to the center of the protein is the RecQ conserved domain characteristic of members of the RecQ family of DNA helicases.[15] The more distally located other region; designated Helicase and RNase D C-terminal domain (HRDC) is present in some RecQ helicases

---

[1] Unless otherwise indicated, WRN refers throughout this chapter to the human protein.

but also in RNase D and its eukaryotic homologs as well as in *Mycobacterium leprae* UvrD helicase.[14,15] The functional importance of HRDC is unknown, as it is missing from both RecQ-like helicases and RNase D in several species. HRDC appears, therefore, to be dispensable for either activity.

Structure of the WRN Promoter

The sequence of a >1.0 kb *WRN* promoter region presents two characteristics of housekeeping genes: greater than 50% (G+C) content and a CpG/GpC ratio higher than 60%.[22] Computer programs prediction and RNase protection assays reveal two transcription start sites within the *WRN* promoter region. No classical TATA or CAAT box is discerned within a stretch of 826 bp in the promoter. This region contains multiple Sp1 binding site sequences and binding sequences for the nuclear factor ETF that specifically stimulates transcription from TATA-less promoters.[22]

Homology of WRN to Other Members of the RecQ Family

The RecQ family of DNA helicases has members present in every examined prokaryote, unicellular eukaryote or vertebrate. All the RecQ proteins have a conserved ~450 amino acid long central domain comprised of seven motifs that include an ATP binding sequence (a Walker A-box) and a DExH box shared by all the RecQ proteins. Homology between WRN and *E. coli* RecQ helicase (610 amino acids, ref. 23), *Saccharomyces cerevisiae* Sgs1 (1447 amino acids, ref. 24), and *Schizosaccharomyces pombe* Rqh1 (1328 amino acids, ref. 25) is confined to the helicase domain. These proteins are therefore considered orthologs rather than homologs of WRN. By contrast, with homology spanning their full sequence, the *Xenopus laevis* FFA-1 (1367 amino acids, ref. 26) and the mouse Wrn (1401 amino acids, ref. 27) proteins are authentic homologs of WRN. Most notably, whereas *E. coli* RecQ, *S. cerevisiae* Sgs1 and *S. pombe* Rqh1 are devoid of the exonuclease domain, it is present in *X. laevis* FFA-1 and mouse Wrn.

Five RecQ-like DNA helicases were identified to date in human cells. The genes encoding these proteins, schematically represented in Figure 2, are *WRN*,[8] *BLM*,[28] *RecQL*,[29,30] *RecQ4*[31,32] and *RecQ5*.[31,33] Mutations in three of these genes result in hereditary genomic instability disorders characterized by predisposition to malignancies and/or early senescence. WS, Bloom syndrome (BS) and Rothmund-Thomson syndrome (RTS) arise from autosomal recessive mutations in *WRN*,[8,9] *BLM*[28,34] and *RecQL4*,[35,36] respectively, (for review see 37). No disease has been associated with mutations in either *RecQL* or *RecQ5*. As illustrated in Figure 2, *WRN* is distinguished from the other four human RecQ helicase genes in that it encodes an N-terminal exonuclease domain in addition to the centrally located RecQ helicase activity.

The WRN Helicase Activity

Isolation of Recombinant WRN Protein

Following the positional cloning of *WRN*,[8] many attempts were made to express the gene in a variety of systems in order to demonstrate that it encodes an active DNA helicase. Success was attained when WRN cDNA was subcloned into baculovirus transfer vectors and expressed in cells of the insect *Spodoptera frugiperda*.[10,11] Histidine-tagged 163-170 kDa recombinant WRN protein purified by Ni^{2+}-NTA affinity column chromatography, possesses a DNA-dependent γ-ATPase activity.[10,11] When challenged by a partial DNA duplex with 5' and 3' recessed ends (10) or by a hybrid of circular single-stranded bacteriophage M13 DNA with a complementary DNA oligomer,[10,11] WRN acts as a DNA helicase by catalyzing dis-

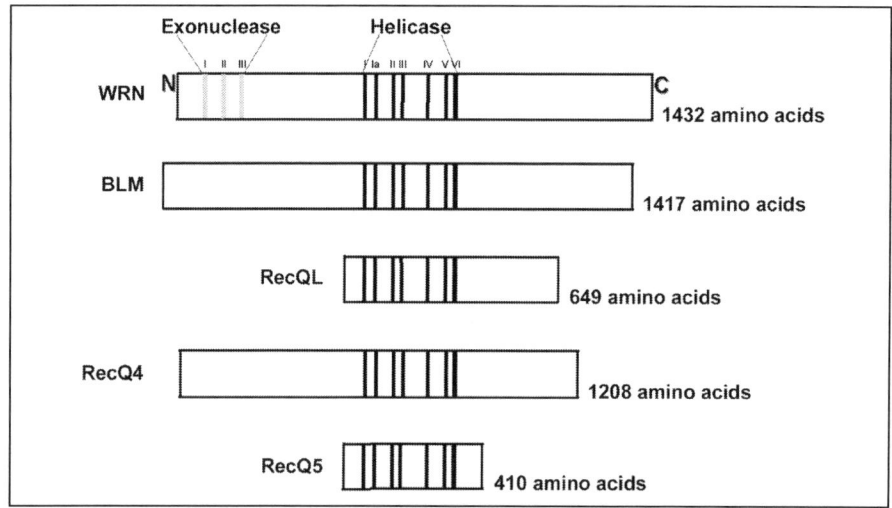

Figure 2. Human RecQ helicases. The five RecQ homologs identified in human cells are schematically represented and aligned by their conserved helicase domain (motifs I-VI). Note that WRN is distinguished from the other four helicases by having an N-terminal 3-motif exonuclease domain.

placement of the oligomer in ATP-dependent reaction (see Fig. 3 for schemes of the DNA substrates). Strand displacement is supported by either ATP (K_m value of 51 μM) or dATP (K_m=119 μM) and to a lesser extent by CTP (K_m=2.1 mM) and dCTP (K_m=3.9 mM).[38] On the other hand, GTP, dGTP, UTP or dTTP fail to effectively substitute for ATP in supporting DNA unwinding by WRN.[38] That the observed helicase activity is intrinsic to the WRN protein is confirmed by the failure of a WRN K577M mutant protein to perform DNA strand displacement.[10] The lysine-to-methionine substitution at residue 577 in this mutant protein abolishes the ATPase activity of motif I (Walker ATPase A site) within the helicase domain.

Properties of the WRN Helicase

WRN is innately a relatively weak DNA helicase, displacing a shorter (e.g., 24-mer) DNA fragment from its hybrid with M13 DNA at a higher efficiency than a longer (e.g., 40-mer) oligonucleotide.[10,11] However, as described elsewhere in this volume, upon its association with replication protein-A (RPA) WRN becomes capable of unwinding longer DNA duplexes. When presented with double-stranded DNA molecules, WRN disrupts most proficiently a DNA duplex with a 3' single-strand tail (Fig. 3), while normally failing to resolve blunt-ended double-stranded DNA.[10] WRN also displaces an 18-mer oligoribonucleotide from its RNA-DNA heteroduplex with M13 DNA, though at a lower efficiency than that of a similar DNA-DNA duplex.[11]

To determine the polarity of the WRN helicase, linear single-stranded M13 DNA is annealed at its 3' and 5'-ends to labeled complementary oligomers of different lengths such that blunt-ended terminal duplexes are formed. Whereas WRN releases the oligomer from the 5'-terminus, it fails to displace the oligomer from the 3'-end, indicating that it unwinds DNA in a 3'→5' polarity[11,38] (helicase directionality is defined as the polarity of its translocation along the enzyme-bound strand). Thus, similarly to *E. coli* RecQ (39), WRN unwinds DNA in a 3'→5' direction.

WRN Helicase Preferentially Unwinds DNA Substrates that Have Alternate Structure

Studies of the DNA substrate specificity of WRN helicase indicate its preference for the unwinding of alternate DNA structures. As detailed below, this proclivity to disrupt unusual DNA structures could point to roles of WRN helicase in resolving topological or aberrant DNA structures that might be formed during the metabolism of genomic DNA. It should be noted in this context that the 3'→5' exonuclease activity of WRN is also marked by preference for DNA substrates that assume alternate structures (vide infra). Two cautionary notes should be raised in considering our current knowledge of the DNA substrate specificities of WRN:

1. All the existing data are based on results of in vitro experiments that employ highly purified recombinant WRN and utilize synthetic oligomer models of various DNA structures. It is possible that modification of the reaction conditions or use of variant DNA sequences or structures might reveal additional or different substrate specificities of the helicase.
2. As described at length elsewhere in this volume, WRN is known to associate with a variety of DNA interacting proteins, some of which modulate the helicase or exonuclease activities. It is plausible that WRN might function in vivo as a component of multiprotein complexes that conduct different DNA transactions. Conceivably, therefore, the interaction of WRN with its associated proteins might modify the inherent substrate preferences of the helicase.

Following is a survey of alternate DNA structures that the WRN helicase utilizes as preferred substrates.

Tetraplex DNA

The first unusual DNA structures that were shown to serve as preferred substrates for the WRN helicase are some types of tetrahelical DNA (also designated tetraplex or quadruplex DNA). DNA sequences that include clusters of contiguous guanine residues readily assemble in vitro under physiological-like conditions into tetrahelical formations. Such four-stranded structures are maintained by guanine-guanine Hoogsteen hydrogen bonds and require Na^+ or K^+ ions for their formation and stability.[40-43] Tetraplexes of different stoichiometries and strand orientation are formed depending on the number of adjacent guanine clusters along a DNA stretch, on the cation present and on the conditions of the reaction. The three main tetraplexes types, schematically depicted in Figure 3, are G'4 unimolecular anti-parallel tetrahelices, G'2 bimolecular anti-parallel variants and G4 four-molecular parallel-stranded species. Although direct evidence for the existence of tetraplex DNA in vivo is still missing, indirect data suggest that tetrahelical structures of various guanine-rich tracts in genomic DNA may contribute to a variety of physiological and pathological DNA transactions. For instance, tetraplex DNA structures were proposed to play roles in meiotic recombination[41,43] and in controlling telomere elongation and stability.[44,45] Also, by arresting the progression of DNA polymerases,[46] tetraplex structures of the d(CGG) trinucleotide repeat sequence might promote expansion of this sequence and engender fragile X syndrome.[47]

Highly purified recombinant WRN has been shown to unwind G'2 bimolecular tetraplex structures of $d(CGG)_7$ that have single-strand tails at their 3' or 5'-end but it fails to resolve blunt-ended G'2 $d(CGG)_7$.[38] WRN disrupts 3'-tailed G'2 $d(CGG)_7$ at efficiencies that are up to 3.5-fold greater than that of partial DNA duplex.[48] In contrast to its ability to unwind tailed G'2 $d(CGG)_7$, WRN was initially reported to be unable to resolve a G'2 tetraplex form of the telomeric DNA sequence $d(TTAGGG)_2$ that has a 8 and 3 nucleotide-long single stand tails at its 5' and 3' ends, respectively. Similarly, WRN was found to be incapable of disrupting a G4 structure of the IgG switch sequence d(TACAGGGGAGCTGGGGTAGA) that has 4 unpaired nucleotides at each of its 3' and 5' ends.[48] In a subsequent study WRN was found to be capable of proficient unwinding of tetrahelical forms of an immunoglobulin switch region sequence and of *Oxytricha* telomeric repeat sequence, both having 7 nucleotide long single-strand

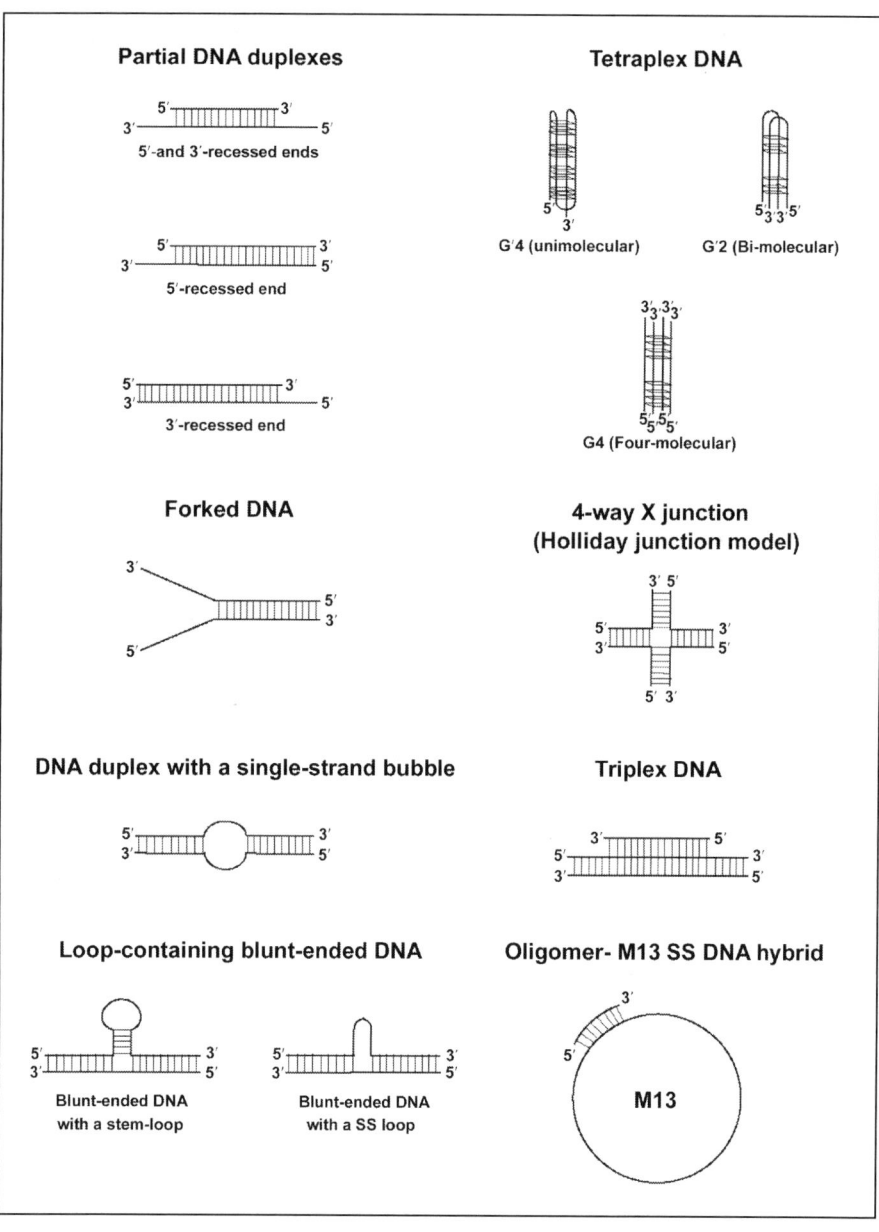

Figure 3. Alternate DNA constructs that serve as in vitro substrates for the WRN helicase or 3'→5' exonuclease. Synthetic DNA oligomers with defined nucleotide sequence are used to construct alternate DNA structures that serve as models for DNA formations that might arise during different in vivo DNA transactions. Forked DNA, partial duplexes and bubble-containing duplex might serve as models for replication intermediates; 4-way X junction (Holliday junction), triplex and tetraplex DNA potentially stand for recombination intermediates; and DNA duplexes that contain a single-stranded bubble or loop or stem-loop might correspond to structures processed during DNA repair. **Partial DNA duplex**: Annealing of an oligonucleotide to a shorter complementary oligomer forms this structure. Illustrated are variant

Figure 3 legend continued next page

Figure 3. Legend continued from last page.
structures that have protruding 5'- or 3'-ends or both. The thicker rods represent DNA strands and the connecting thin bars denote Watson-Crick hydrogen bonds between complementary nucleotides. **Forked DNA**: A model forked DNA structure is formed by two annealed oligomers that contain complementary nucleotide tracts only along their respective 3' and 5'-end segments. **DNA duplex with a single-stranded bubble**: To generate this structure, two complementary oligomers are annealed that contain a non-complementary tract in their respective centers. Variant structures might differ by the position and size of the single-stranded bubble. **Loop-containing blunt-ended DNA**: To generate duplex DNA with a single-stranded loop in one of the two strands, an oligomer is annealed to a partially complementary DNA fragment that includes an extra non-complementary central tract that loops out. To form duplex DNA with a stem-loop in one strand, the annealed non-complementary stretch has a short self-complementary region that becomes a stem. **Tetraplex DNA**: oligomers that contain tracts of contiguous guanine residues are capable of forming non-Watson-Crick hydrogen bonded guanine quartets that stabilize tetraplex DNA structures. The different number of clustered guanine residues and the quartets that they form as sketched in the different tetraplex structures, arbitrarily signify different arrangements of guanine residues along the tetraplex-forming DNA strands. Schematically depicted are three variants of tetraplex DNA: G'4 unimolecular anti-parallel unimolecular structure; G'2 bimolecular anti-parallel formation; and G4 four-molecular parallel-stranded structure. **4-way X junction**: This structure is constructed from four oligomers, each complementing segments in two other oligomers. **Triplex DNA**: A three-stranded DNA structure can be composed of a pyrimidine-rich strand that pairs by Watson-Crick bonds with a complementary purine-rich strand and with another pyrimidine-rich strand by non-Watson-Crick bonds. Variant triplexes differ by the arrangement of purine- and pyrimidine-rich strands and the position and length of their single-stranded segments. **Oligomer-M13 single-strand DNA hybrid**: This structure is constructed by annealing a short synthetic oligomer to a selected complementary sequence in bacteriophage M13 circular single-stranded DNA.

tails at their 3'-ends.[49] These observations suggest that the ability of WRN to unwind a tetraplex DNA substrate might be determined by the length and orientation of the single-strand tail that it has rather than by its nucleotide sequence or structure. In accord with the earlier report,[48] it was also shown that whereas WRN fails to resolve a blunt-ended tetraplex, it unwinds the tailed tetraplex DNA structures at efficiencies that exceed those of any other tested DNA substrate.[49] The capacity of WRN to efficiently unwind tetraplex DNA structures is shared by other members of the RecQ helicase family, human BLM[49,50] and *S. cerevisiae* Sgs1.[51]

What in vivo function might WRN serve in unwinding tetraplex DNA structures? At this time one can only conjecture on this matter, basing speculations on some suggestive pieces of evidence.

First, by resolving aberrant tetraplex DNA structures WRN might clear the way for DNA polymerases during replication or repair synthesis. Evidence shows that the progression in vitro of various DNA polymerases, including the eukaryotic replicative polymerases α, δ and ε, is blocked by hairpin and G'2 tetraplex structures of a $d(CGG)_n$ sequence in template DNA.[46] Functional[52] and presumably physical[53] association of polymerase δ with WRN, permits coordinate unwinding of the hairpin and tetraplex structures by the helicase and enables traversal of the unfolded $d(CGG)_n$ tract by the polymerase.[46] Polymerases α or ε do not form a functional complex with WRN and the helicase is not effective in alleviating their blocking at $d(CGG)_n$ hairpin and tetraplex template structures.[46] It appears, therefore, that conjoined action of WRN and a DNA polymerase that enables concerted unwinding of template secondary structures and progression of the polymerase, is dependent on the formation of a complex between the two enzymes.

Second, by resolving tetraplex structures of telomeric DNA, WRN might play a role in telomere metabolism. Tetrahelical structures of telomeric DNA have been implicated in the determination of telomere length and stability.[44,45] If WRN partakes in the regulated unwinding of tetraplex domains in telomeric DNA, its absence in WS cells should be reflected by aberrant telomere metabolism. Indeed, several lines of evidence point to telomere anomalies in WS cells. It has been reported that although telomeres of WS skin fibroblasts shorten at an accelerated rate, the cells cease to divide when the mean length of their telomeres is greater than in normal senescent cells.[54] Telomeres are also trimmed at an accelerated rate in Epstein-Barr Virus (EBV)-transformed WS lymphoblastoid cells but at the time of their exit from the cell cycle these cells have a wider range of telomere lengths than do their normal counterparts.[55] Ectopic expression of human telomerase in WS cell lines extends their life span and culminates in their immortalization.[56-58] However, the patterns of mRNA expression in telomerase-immortalized WS and normal cells differ substantially, suggesting that telomerase expression does not faithfully reverse the destabilized WS genotype.[58] Additional evidence that links WRN to telomere metabolism is provided by the co-localization of WRN and the telomere repeat binding factors TRF1 and TRF2 in nuclear foci of immortalized human cells that lack telomerase.[59] Analogies are drawn between the proposed WRN-telomerase connection and the requirement for Sgs1 activity in the maintenance of telomeres in telomerase-deficient yeast cells.[59] It remains to be explored whether or not the observed association of WRN with telomeres and the anomalies of telomere metabolism in WS cells reflect a direct role for WRN in a regulated unwinding of tetraplex telomeric DNA.

Third, resolution of tetraplex DNA structures by WRN might play a part in DNA recombination. The formation of parallel-stranded four-molecular tetraplex structures has been suggested to be required for the alignment of homolog chromosomes during meiosis.[41,43] WRN preferentially unwinds G4 DNA[48,49] and tri-substituted acridine derivatives that are potent and selective ligands for tetraplex DNA inhibit the unwinding of a G4 form of an immunoglobulin switch region sequence.[60] This unwinding activity might be instrumental in the resolution of tetraplex recombination intermediates. Bloom syndrome is marked by hyper-recombination between sister chromatids and homologous chromosomes. The disruption of G4 DNA by BLM helicase was proposed to reflect its function in resolving tetrahelical recombination complexes.[50] By analogy, a reported aberrant mitotic recombination in WS cells,[61] might signal a similar or complementary role for WRN in the unwinding of tetraplex recombination complexes that form during mitosis.

Holliday Junction (4-Way X Junction)

Four-way Holliday junctions formed between paired sister chromatids and homologous chromosomes are processed to yield recombinant DNA products. Disruption by DNA helicases of such structures and prevention of their subsequent processing might suppress hyper-recombination. BLM helicase selectively binds in vitro 4-way X junction synthetic DNA molecules that serve as models for Holliday junction (see Fig. 3 for a scheme).[62] BLM also promotes ATP-dependent DNA branch migration on Holliday junction-containing a structures prepared by RecA-mediated strand exchange.[62] Concordantly, BLM unwinds in vitro a synthetic blunt-ended 4-way X junction DNA substrate at an efficiency that is second only to G4 DNA and is 10-fold higher than that of the standard 3'-tailed duplex DNA standard helicase substrate.[49] These results were interpreted as indicative of a role for BLM in the suppression of hyper-recombination.[62] Very similar data were gathered on the proclivity of WRN to disrupt Holliday junction structure and to promote DNA branch migration. Purified recombinant WRN unwinds in vitro a 4-way X junction DNA molecule at a relatively high rate that is exceeded only by the rates of unwinding of G4 tetraplex DNA and of bubble-containing duplex DNA.[49] Resembling BLM, WRN is also capable of promoting branch migration on a Holliday junction-containing α structures DNA substrate.[63] These capacities of WRN were interpreted as reflecting its potential role in the in vivo dissociation of recombination intermediates and the processing of potentially recombinogenic broken replication forks. WRN is proposed to act as an anti-recombinase that affords functions required for genomic stabilization during DNA metabolism.[63] Notably, however, the distinct phenotypic differences between BS and WS are not reflected in the closely similar capacities of BLM and WRN to resolve in vitro 4-way X junctions and to promote branch migration. Whereas hyper-recombination is a hallmark of BS cells, it is not a prominent characteristic of cells derived from WS patients. Assumingly, therefore, the functions provided by these two helicases are overlapping but non-identical. In vitro experiments that utilize purified BLM or WRN proteins by themselves are not sufficient to reveal subtle functional differences between these helicases. WRN or BLM are unlikely to act in vivo as solitary enzymes but rather, might probably function in concert with specific interacting proteins. As described elsewhere in this volume, various proteins form complexes with WRN and modulate its activity. Hence, reconstituted WRN-containing multiprotein complexes might mirror the in vivo situation more faithfully, providing a better mechanistic description of the handling of recombination intermediates by WRN and delineating its distinction from BLM.

Triplex DNA

Triple-helical DNA (triplex DNA) structures are formed by the insertion of a third strand into the major groove of duplex DNA. Such DNA structures, recognized for many years (see ref. 64 for a review), are generated most readily when either a polypurine or a polypyrimidine strand is placed within a polypurine · polypyrimidine duplex. Triplex DNA can arise when one strand of an appropriate duplex partially melts and folds back to form a triple helix with the remaining duplex.[64] Some lines of evidence suggest that triple helical DNA exists in vivo. First, triplex DNA-specific antibodies bind to nuclear DNA and partially inhibit replication and transcription.[65] Second, triplex DNA binding proteins are present in human cells[66,67] and in *S. cerevisiae*.[68]

Triplex DNA has been identified as another alternate DNA structure that serves as a substrate for WRN and BLM helicases.[69] Both enzymes similarly unwind triple helical DNA in ATP-dependent reaction, provided that the third strand ends in a 3' single-strand tail. By contrast, the two helicases do not require the presence of a duplex extension that forms a fork at the point of junction between tail and triplex.[69] It is not readily apparent what possible in vivo function the triplex DNA disrupting activity of WRN might serve. The reported recombinogenic capacity of triple helical DNA structures[70,71] and their ability to arrest Holliday junction progression in vitro[72] might suggest that triplex DNA contributes to genomic instability. If this is the case, than by disrupting triple-helical structures in genomic DNA, WRN (and BLM) could act in vivo to reduce the accumulation of aberrant structures in genomic DNA.

Bubble-Containing Duplex DNA

With 3'-tailed duplex DNA being the standard substrate to measure WRN activity,[10,11,49] variant structures of double-stranded DNA were examined for their potency to serve as substrates for the helicase. Unlike *E. coli* RecQ that is capable of initiating DNA unwinding from a blunt end of duplex DNA,[73] WRN cannot disrupt blunt-ended DNA.[10,49] WRN also fails to unwind at a measurable rate a DNA duplex that ends with a 5' single-strand stretch (Fig. 3) or a nick-containing duplex.[49] However, whereas the helicase is also unable to disrupt a DNA duplex that contains at its center a 4-nucleotide single-stranded bubble (see Fig. 3 for a scheme of bubble-containing DNA), it efficiently unwinds a double-stranded molecule that has a 12-nucleotide bubble.[49] In fact, the 12-nucleotide bubble-containing DNA substrate is disrupted at a rate that is second to only to that of G4 DNA and that exceeds the rate of unwinding of 3'-tail duplex DNA by more than 10-fold.[49] WRN similarly unwinds in ATP-dependent fashion an 80-nucleotide long blunt-ended duplex that contains a centrally located bubble of

21 nucleotides.[75] The preferential unwinding of duplex DNA with a single-stranded bubble is matched by the efficient binding of WRN to this substrate. Whereas WRN associates with bubble-containing duplex DNA in ATP-dependent reaction, it does not detectably bind duplexes with blunt or 3'-recessed ends.[74] Notably, bubble-containing duplex DNA also serves as a preferred substrate for the 3'→5' exonuclease function of WRN (refs. 74,75, vide infra). The capacity of WRN to resolve a bubble containing blunt-ended duplex as well as its ability to unwind the blunt-ended X-junction DNA substrate (see above), indicate that a single-stranded 3'-tail is not obligatory for the helicase activity. Hence, WRN does not require a 3'-single-strand tract to load onto and to track along the substrate molecule until it encounters a duplex DNA region.

It is difficult to construe at this point in time the in vivo significance of the preferential activity of WRN helicase (and exonuclease) with bubble-containing duplex DNA. One is tempted to speculate that a single-stranded bubble amidst duplex DNA emulates structures formed during DNA metabolism in vivo such as a bubble-like origin of bi-directional replication. To advance our understanding of the special affinity of WRN helicase for bubble-containing DNA substrates, more data need to be gathered on their interaction. The DNA substrate preference of WRN should be examined, therefore, under different reactions conditions, with diverse bubble-containing DNA constructs and particularly, upon association of WRN with its interacting proteins.

Forked DNA

Forked duplexes (see scheme in Fig. 3) are also unwound by WRN helicase. These synthetic model DNA molecules are disrupted by WRN at about 3-fold the rate of unwinding of 3'-tailed duplex DNA.[49] Interestingly, BLM also resolves this DNA substrate at a somewhat higher efficiency than WRN.[49] Data indicate that WRN (and BLM) prefer structures such as G4 and X junction DNA that possibly emulate recombination intermediates, to the forked duplex DNA molecules that might represent branched replication forks.[49] At first glance these results suggest a preference of WRN and BLM for recombinogenic rather than replicative DNA structures. However, one should bear in mind the limited variety of DNA substrates that were tried to date, the narrow range of assay conditions attempted and the absence of WRN-interacting proteins in the reaction mixtures. Hence, as with other DNA substrates, conclusive determination of the in vivo relevance of the disruption of forked DNA by WRN awaits the accumulation of more experimental data.

Binding of WRN to DNA

Studies on the DNA binding properties of WRN were conducted in parallel to investigations of its DNA unwinding activity. In one report WRN was found to be unable to bind synthetic DNA duplexes with blunt or 3'-recessed ends.[74] By contrast, this helicase-exonuclease forms stable complexes with synthetic duplex DNA that contains a single-strand bubble.[74,75] In a different study, WRN was shown to associate with single-stranded circular pKS plasmid DNA with a binding constant of 10^8-10^9 M^{-1}.[76] Notably, the binding constant for double-stranded circular pKS DNA is more than 5-fold lower than that for single-stranded DNA.[76] The results of both studies point to single-strand DNA and to single-strand bubble within duplex DNA as targets for the binding of WRN. However, whereas in these two reports WRN fails to bind synthetic duplex DNA to measurable levels,[74,75] others found that it does associate to a low but significant degree with double-stranded plasmid DNA.[76] This discrepancy could be due to the use of different types of duplex DNA in the different studies or to the inclusion of ATP in the DNA binding reaction mixture in one report[74] but not in the other.[76]

In light of the overt sensitivity of WS cells to the DNA damaging agent 4-nitroquinoline-1-oxide (4-NQO)[77,78] binding of WRN to plasmid DNA that contains

4-NQO-induced adducts was compared to that undamaged DNA. WRN was found to bind at equal affinities undamaged DNA, DNA damaged by 4-NQO or DNA exposed to UV radiation or to methylene blue and light.[76] Taken together, these results indicate that within the limits of detection of the assay used, WRN does not appear to interact preferentially with DNA that accumulates mutagenic insults.

Inhibitors of WRN Helicase Activity

Several agents potently inhibit the helicase activity of WRN protein. However, no agent has yet been discovered that either exclusively blocks the WRN helicase activity without affecting other RecQ-type helicases or that specifically inhibits the unwinding of selected DNA substrate(s). A number of DNA binding compounds hinder the disruption of a partial DNA duplex by WRN and BLM. The chemically closely similar minor groove binding drugs distamycin A and netropsin block most potently and to a similar extent WRN or BLM-catalyzed displacement of an oligomer from its partial duplex with M13 DNA.[79] Inhibition by distamycin A is highest on a DNA substrate that contains a five A-T base pairs tract (K_i ~0.5 µM), whereas netropsin exerts maximum inhibition on the unwinding of a substrate that has four A-T pairs (K_i ~1 µM).[79] This dependence on the length of the A-T stretch matches the different sequence preference of the two compounds: whereas distamycin A maximally binds to DNA duplex that contains tracts of five A-T base pairs, netropsin binds best to DNA with a four A-T tract. Moreover, the apparent K_i values of the two drugs are comparable to the K_d values of their association with their preferred DNA substrates. Although these results indicate that the two drugs act through their binding to DNA rather than to the helicases, both are very poor inhibitors of UvrD helicase (K_i >100 µM).[79] Hence, the interaction of distamycin A and netropsin with DNA appears to differently affect different helicases. This disparate interaction of drug-bound DNA with diverse helicases is delineated by the contrast between the moderate effectiveness of inhibition of WRN and BLM by mitoxantrone (K_i ~10 µM and >10 µM, respectively) and the potent blocking of UvrD helicase by this drug (K_i ~1 µM).[79] Distamycin A and netropsin as well as a number of other less effective inhibitors, are not selective for WRN as they all block the activity of BLM to a very similar extent.[79]

Tri-substituted acridine derivatives that bind tetraplex DNA tightly and selectively inhibit the unwinding by WRN and BLM of a G4 DNA form of an immunoglobulin switch region sequence.[60] However, in addition to blocking the disruption of G4 DNA, these compounds also inhibit with similar potency the unwinding by WRN and BLM of forked duplex DNA.[60] Thus, both drugs do not discriminate between the two tested human RecQ helicases and are similarly effective with G4 and B-form DNA molecules. With tetraplex DNA being a preferred substrate for WRN helicase,[48,49] it will be interesting to examine the potency and selectivity of the inhibition of its unwinding by a number of additional, recently discovered tetraplex DNA interacting drugs.[80-83]

The WRN 3'→5' Exonuclease

Detection of an Exonuclease Activity As an Integral Part of the WRN Protein

Computer analysis of the amino acid sequence of the N-terminal portion of WRN protein provided the first clue that in addition to a DNA helicase activity it also contains an integral exonuclease. Iterative strategy utilizing several computer methods[14,15] and Hidden Markov Model analysis[16] discovered a three-motif conserved nuclease domain in the N-terminal region (residues 78 to 219) of WRN. Employing biochemical, immunochemical and molecular genetic methods, several groups showed shortly thereafter that the N-terminal segment of WRN catalyzes exonucleolytic degradation of DNA.[17-19,84] Early on, purified histidine-tagged recombinant WRN protein was shown to digest DNA exonucleolytically in a 3'→5' direction.[17-19] This polarity of the exonuclease was later validated in numerous studies.[74,85-90] However, a contrasting study found that WRN degrades DNA exonucleolytically with a 5'→3' polarity.[84] The basis for this discrepant result is still unclear. Interestingly, an expressed minimal N-terminal exonuclease domain of WRN that encompasses amino acids 70-240 was recently found to contain in addition to 3'→5' exonuclease an endonucleolytic activity that hydrolyzes a 5'-protruding strand.[90] It is not clear whether the full-length WRN protein also possesses this activity and if it is related in any way to the reported 5'→3' exonuclease activity.

Several lines of evidence confirm that the exonuclease activity is a covalently linked element of the WRN polypeptide and that the helicase and exonuclease reside in different domains of this protein. First, the 3'→5' exonuclease activity co-purifies with the helicase and ATPase activities through successive steps of ion exchange and affinity chromatography to yield an apparently homogeneous 160-kDa protein that possesses all three activities.[17] Second, active exonuclease activity is maintained by purified recombinant WRN mutant proteins that are devoid of helicase activity due to amino acid substitutions in the helicase domain. Reciprocally, mutant WRN proteins with a truncated N-terminal domain or that have amino acids substitution in this region, lack exonuclease activity but sustain their helicase activity.[17,19] In line with this finding, an expressed cloned minimal fragment of *WRN* that encodes a polypeptide encompassing residues 70-240 also has a 3'→5' exonuclease activity but lacks helicase activity.[90] Third, antiserum directed against the WRN protein specifically co-precipitates both the helicase and exonuclease activities.[17] Put together, these results substantiate the unique status of WRN among other RecQ-like helicases as the only protein that hosts within the same polypeptide both an exonuclease and a prototypic helicase.

WRN 3'→5' Exonuclease Selectively Digests a 3'-Recessed Strand in Duplex DNA and Requires ATP for Maximum Activity

The WRN 3'→5' exonuclease differs from most other known exonucleases by the distinctive set of DNA substrates that it is able to digest. Highly purified recombinant WRN generates 5'-dNMP products in degrading a 3'-recessed strand in a partial DNA duplex (ref. 18, see Fig. 3 for an illustration of 3'-recessed DNA substrates).[18] Similar nucleolytic activity is found with a 3'-recessed DNA-RNA heteroduplex.[87] WRN can also initiate a nucleolytic attack on a 3'-terminus from a 12-nucleotide long nick in double-stranded DNA.[87] Recessed DNA strands that terminate with either a 3'-OH or 3'-PO$_4$ group are degraded to a similar extent by the exonuclease. Also, a recessed DNA strand that has a single 3'-terminal mismatched nucleotide is digested more efficiently than a fully complementary strand.[18] In contrast to the efficiently digested 3' recessed strands, WRN fails to detectably hydrolyze single-stranded DNA, blunt-ended double-stranded DNA, a 3'-protruding strand in a partial duplex or a 3'-recessed strand with two 3'-mismatched bases.[18] Purified WRN differs, therefore, from most exonucleases by its stringent preference for 3'-recessed strands in duplex molecules and by its failure to hydrolyze single-stranded DNA or a protruding 3'-end single-strand. This distinctive substrate specificity of human WRN is underlined by the finding of similar DNA substrate preference of a highly purified preparation of mouse Wrn.[87] It should be noted, however, that the narrow range of substrates that purified WRN exonuclease attacks might well be broadened upon its association with auxiliary proteins. Some examples already exist for a mutual modulation of activity in complexes of ancillary proteins with WRN exonuclease. For instance, as discussed elsewhere in this volume, the association of Ku antigen with WRN greatly enhances the processivity of the exonuclease[85,86] and allows it to efficiently digest blunt-ended DNA and a 3'-protruding strand.[86] Another case of functional interaction between WRN exonuclease and an associated protein is that of p53. Binding of p53 to WRN results in repression of its exonuclease activity,[91] and in the enhancement of p53-dependent transcriptional activity and induction of p21 *Waf1*.[92] It appears, therefore, that more information on the effect of auxiliary proteins on WRN activity needs to be gathered before definitive conclusions on the substrate preference of the exonuclease are drawn.

WRN exonuclease is also distinguished from most other DNA exonucleases by its requirement for hydrolysis of a nucleotide triphosphate for maximum activity. Although the exonuclease digests recessed 3'-strands in the absence of ATP or dATP, hydrolysis is stimulated 2 to 4.5-fold in their presence.[18] A more modest enhancement of the exonuclease activity is attained in the presence of CTP, UTP or dTTP. The non-hydrolyzable ATP analog ATPγS does not affect the exonuclease activity, indicating that the nucleotide triphosphate must be hydrolyzed to stimulate the exonuclease.[18]

WRN Exonuclease Preferentially Degrades Alternate Structure DNA Substrates

Similarly to its helicase activity, the WRN exonuclease preferentially degrades DNA substrates that have alternate structures. A systematic survey of the relative degradation of a series of alternate DNA structure revealed a clear preference of the WRN exonuclease for substrates that contain either a bubble or a single-strand loop within duplex DNA (see Fig. 3 for schemes of these substrates). The exonuclease digests in ATP-dependent fashion blunt-ended or partially double-stranded DNA that contains a single-strand bubble at rates that are higher by 2.5-fold than the rate of degradation of a 3'-recessed strand in a partial duplex.[74] A study of the degradation of a 80-nucleotides blunt-ended duplex that contains a 21-nucleotides single-stranded bubble reveals that 50% of this substrate are digested by 100-fold lower amounts of WRN than those required for 50% unwinding of a 3'-recessed partial duplex.[75] Also, the exonuclease is much more processive with this bubble-containing substrate than with the 3'-recessed DNA.[75] DNase I footprinting analysis shows that WRN binds to the bubble region and that its association with the melted DNA stretch is most probably responsible for the exonucleolytic digestion of the nearby ends.[75] Elevated rates of degradation are also obtained with blunt-ended duplex that contains a stem-loop or a single-stranded loop and with 4-way or 3-way X junction DNA.[74]

What might be the in vivo significance for the preference of WRN exonuclease for alternate structures of DNA? Strikingly, both the helicase and exonuclease activities of WRN protein display increased activity with alternate DNA structures. As argued above for the helicase activity of WRN, the processing of unusual DNA structures might reflect important functions that WRN has in the conduct of DNA transactions such as replication, recombination or telomere extension and stabilization. However, to execute different tasks in DNA metabolism, the helicase and exonuclease activities might be differently modulated when WRN forms different complexes with diverse auxiliary proteins. The isolation and characterization of such complexes is thus required in order to fully catalogue the DNA substrate preferences of the WRN exonuclease (and helicase). Also, we still have no knowledge which WRN activity predominates when the protein encounters non-canonical DNA formation. It might be the helicase that unwinds most of such structure or alternatively, the exonuclease activity outweighs the helicase. As detailed below, solving the problem of coordination between the two activities should assist in reaching a better understanding of this problem.

Homo-Oligomerization of the WRN Exonuclease

Most of the known DNA helicases, including BLM,[93] oligomerize to form dimers or hexamers (reviewed in 94). An N-terminal fragment of WRN that includes amino acids 1-333 with a predicted molecular size of 42 kDa, was reported to elute from a Superdex S-200 gel filtration column mostly as a 130 kDa protein. A minor fraction of the eluted fragment behaved as a 250 kDa protein. These results suggest that the 333-amino acid polypeptide that spans the WRN exonuclease domain oligomerizes to mostly form a homo-trimer with a minor proportion of the protein entering into hexamers.[87] Full length WRN protein (molecular mass 165 kDa) was reported to elute as a 465 kDa molecule, indicating that it too forms a trimer.[87] Recent results indicate that a more minimal N-terminal fragment of WRN, spanning amino acids 70-240, also forms a trimer. This recombinant WRN polypeptide (molecular size ~21 kDa) elutes from a Superdex S-200 gel filtration column as a 65 kDa protein.[90] These results are corroborated by atomic force microscopy determination of the molecular volumes of more than 1100 molecules of this N-terminal minimal fragment of WRN. It is found that ~90% of the fragment molecules are in a trimeric form whereas ~10% are in hexamers.[90] Last, atomic force microscopy results show that the addition of 3'-recessed double-stranded DNA tilts the equilibrium between the two oligomeric forms of this N-terminal fragment toward the hexameric mode. Trimers and hexamers of the WRN minimal polypeptide are present at about equal amounts when a 3'-recessed partial duplex DNA is present.[90] It will be interesting to expand these analyses to study formation of oligomeric states of the WRN holoenzyme in the presence of alternate DNA structures that serve as preferred substrates for the helicase and exonuclease.

Coordination of the Helicase and Nuclease Activities

In a study of the modes of action of the WRN helicase and exonuclease with a forked DNA substrate, it was found that the two activities act coordinately by unwinding the DNA at the forked end while exonucleolytically digesting its 3' blunt end.[95] The length of the forked DNA substrate was reported to determine which activity, helicase or exonuclease, predominates.[95] The question of possible interdependence between the helicase and exonuclease activities was addressed by comparing the exonucleolytic activities of a full WRN polypeptide and a truncated mutant WRN protein that contains only the exonuclease domain. Whereas the intact WRN protein binds to and efficiently digests an 80-nucleotide duplex that contains a 21-nucleotide single-stranded bubble, a mutant protein that lacks the helicase domain, neither binds nor digests this substrate.[75] This finding was interpreted as showing that the binding of DNA by WRN and the operation of its helicase and the activity of the exonuclease are coordinated.[75] However, it should be noted that any model arguing for coordination between the two activities of WRN has to confront the problem of their opposite directionalities. As illustrated in Figure 4, the WRN helicase migrates with a 3'→5' direction, as defined by the protein-bound strand whereas the exonuclease proceeds in a 3'→5' polarity, as defined by the free strand. The dilemma is, therefore, how do these two activities act in concert while appearing to move away from each other as the enzymes starts at a DNA gap (Fig. 4). Concordantly, in utilizing a forked DNA molecule, the two activities would appear to migrate toward each other (scheme not shown). One hypothesized mechanism that accommodates these opposing directionalities is illustrated in Figure 4. Gapped or forked DNA substrate is warped such that the helicase and exonuclease domains of the enzyme end up facing two opposite ends of the DNA. Such twisting of the DNA substrate allows the two activities of WRN to advance in the same direction. An alternative possibility is that the large size of the experimentally suggested trimeric or hexameric WRN[87,90] allows it to cover a DNA stretch long enough for the helicase and exonuclease to simultaneously process short tracts at opposite ends of the substrate DNA molecule. Third, migration of the DNA through a static enzyme and alternate presentation of its opposite termini to the respective catalytic sites of the helicase and exonuclease, could also permit coordinate action of the two activities.

The Transcriptional Activation Domain: WRN As a Transcription Activator

Some evidence suggests that the acidic amino acid direct repeat located between the N-terminal exonuclease domain of WRN and its central helicase region (see Fig. 1), might play a role in transcriptional activation. Human WRN protein activates transcription in a yeast

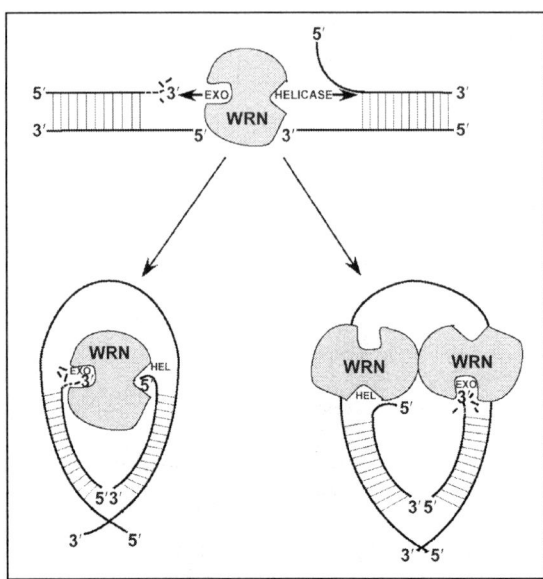

Figure 4. Hypothesized mechanism for coordinate action of the WRN 3'→5' helicase and 3'→5' exonuclease. WRN helicase and exonuclease have opposite polarities: whereas the helicase migrates at 3'→5' direction (defined by its translocation along the enzyme-bound strand), the exonuclease advances with 3'→5' polarity (defined by the attacked enzyme-free strand). As shown in the upper scheme, when faced with a gapped DNA duplex substrate, the helicase and exonuclease appear to translocate away form each other. Hypothetical models shown in the two lower illustrations demonstrate how by flexing around, the DNA presents two opposite termini to the catalytic sites of the helicase and exonuclease. The left hand illustration depicts a monomeric WRN and the right hand cartoon shows a dimeric (or multimeric) enzyme.

system and the WRN transcription activation domain is mapped to the acidic repeat at residues 315 to 403.[96] Mutant WRN protein that lacks the 27 amino acid direct repeat is unable to stimulate transcription, suggesting a direct permissive role for this region in transcription.[97] The possible effect of WRN on transcription is supported by the observation that while permeabilized WS cells show diminished transcription rates, the addition of normal cell extracts to WS cell chromatin restores normal levels of transcription.[98] As detailed elsewhere in this volume, WRN protein localizes to transcriptionally active nucleoli[98-100] and its nuclear-nucleolar trafficking depends on the transcriptional state.[100] These observations might be construed as indications for a role for WRN in the transcription of ribosomal DNA. Evidence to the contrary indicates, however, that steady-state levels of 28S rRNA remain constant during the life span of WS fibroblasts.[101] It is not clear at present whether the acidic repeat domain of WRN is solely responsible for the activation of transcription or rather; the helicase and/or exonuclease also contribute to this activity. Future work should substantiate the role of WRN in transcription and illuminate its mechanism.

Recent Results

While this volume was prepared for publication, some noteworthy results were reported that add to our knowledge of the biochemistry of WRN.

DNA Substrate Utilization by the WRN Helicase/Exonuclease

Requirements for a 3'-single-strand DNA tail for the unwinding of single-stranded tailed duplex DNA structures were assessed by placing a streptavidin-biotin steric block at selected positions along the single-strand tail of the DNA substrates. Whereas blocking the very end of the 3- single strand tail or the single-strand/double-strand junction do not affect duplex DNA unwinding by WRN, the helicase is completely stalled when streptavidin is bound to the 3' single-strand DNA tail 6 nucleotides upstream to the junction.[102] However, the helicase activity is not impeded when streptavidin is similarly bound 6 nucleotides upstream to a junction in a forked version of the DNA substrate. Similar result is obtained with a 5'-flap substrate. These results were interpreted as evidence for a preferential recognition and loading of WRN at single-strand/double-strand junctions in fork/flap double-stranded substrates which function as preferred substrates for WRN.[102]

In a study on the utilization of Holliday junctions as helicase substrates, it was reported that WRN, as well as BLM helicase, cooperate with p53 in processing these DNA structures.[103] Binding of purified recombinant p53 to WRN or BLM helicases inhibits their ability to unwind synthetic Holliday junctions in vitro. However, phosphorylation of p53 at Ser^{376} or Ser^{378} completely abolishes its inhibitory effect. It is suggested that p53 modulates recombinational repair that WRN or BLM mediate and that post-translational modifications of p53 regulate its functional interaction with the two helicases.[103]

A recent report finds a displacement loop DNA (D-DNA) substrate to be a preferred substrate for both the helicase exonuclease functions of WRN. This substrate, that emulates an intermediate generated during the strand invasion step of some recombinational transactions, is preferentially bound and unwound by WRN helicase. Further, the 3' terminus of the invading strand of this DNA structure is readily attacked by the 3'→5' exonuclease activity.[104] These results might suggest a role for WRN in the processing of D-loop structures that are formed during metabolism of cellular DNA.[104]

Regulation by Phosphorylation of WRN Catalytic Activities

DNA-dependent protein kinase (DNA-PK), which forms a complex with Ku antigen, is found to phosphorylate the WRN protein in a Ku-dependent manner.[105] As a result of this in vitro phosphorylation of WRN, its 3'→5' exonuclease activity is inhibited. Conversely, removal of the phosphate groups by Ser/Thr phosphatase enhances both the helicase and exonuclease activities of WRN. Accordingly, exposure of cells to DNA damaging agents results in phosphorylation in vivo of WRN protein.[105] It appears, therefore, that WRN protein is a target for the action of DNA-PK and that phosphorylation might regulate both its helicase and exonuclease activities.

Summary

WRN occupies a distinct position among most of the known DNA helicases or exonucleases in being a bi-functional enzyme, hosting in a single polypeptide two distinct catalytic sites—one executing unwinding of multi-stranded DNA and another digesting DNA exonucleolytically. In possessing these two activities, WRN is distinguished from all the known members of the RecQ helicase family that are present in prokaryotes, unicellular eukaryotes, or vertebrates – including human cells. As argued elsewhere,[100] in considering the possible in vivo functions of WRN, we still do not know whether it is the helicase or the exonuclease function or both whose absence in WS cells causes the distinctive pathology of the syndrome. Hence, a detailed study of both activities is vitally needed for the elucidation of the molecular etiology of WS.

The expression and purification of recombinant WRN protein enabled the gathering of the existing wealth of information on the enzymology of both the helicase and exonuclease. As was described in this chapter, each activity is distinguished from other known helicases or exonucleases by several criteria. However, it is the proclivity of both activities to preferentially process non-canonical DNA structures that is arguably, their most distinct feature. Tetraplex DNA, bubble-containing duplex DNA or Holliday junction structures are disrupted by the purified helicase at higher efficiencies than 3'-tailed partial duplex. Concordantly, DNA duplexes that contain single-stranded bubble, loop or stem-loop are digested by the exonuclease more proficiently than 3'-tailed partial duplex. As compellingly argued elsewhere,[101] this preference for alternate DNA structure is highly suggestive of roles that the WRN helicase and/or exonuclease might play in resolving aberrant structures in genomic DNA. Such unusual formations of DNA can form in the course of replication, recombination or repair. Construction and use of DNA structures that better approximate DNA formations that exist in vivo should expand our understanding of the DNA substrate preference of the helicase and exonuclease.

To elucidate the relative importance of the WRN helicase and exonuclease in DNA metabolism and in the molecular mechanism of WS, a better understanding of the coordination between these two activities must be attained. As described above, we still await an explanation of how the two activities operate in concert despite their opposite directionalities. However, it might also be that with certain DNA substrates and upon association of WRN with auxiliary proteins, only one activity predominates. Thus, to achieve a better understanding of the coordination between the helicase and exonuclease activities, they should be assayed under a wide variety of reaction conditions and with diverse substrate.

All considered, it must be pointed out that rapidly gathering evidence indicates that WRN operates in conjunction with multiple other proteins that also participate in various aspects of DNA metabolism. As detailed elsewhere in this volume and reviewed in refs. 100 and 101, WRN protein associates physically and/or functionally with proteins such as DNA polymerase δ, replication protein A (RPA), Ku antigen, proliferating cell nuclear antigen (PCNA), p53, 5'-flap endonuclease 1 (FEN-1), and topoisomerases I and II. The association of WRN with some of these proteins modulates the levels of activity and substrate specificities of the helicase or exonuclease. Reciprocally, WRN also affects the activities of some of its interacting proteins. These observations strongly argue that WRN operates in vivo in complex with other proteins. It is tempting to speculate that WRN constitutes a component of diverse multi-protein complexes that perform different DNA transactions. Modulation of the WRN activities by specific partner proteins in any specific supra-molecular complex should then determine the relative weight of the helicase and exonuclease and affect their substrate DNA specificities. Hence, a most promising direction of research is the study of properties of the WRN helicase and exonuclease when they form complexes with ancillary proteins. Results obtained by such investigations should bring us closer to an understanding of the in vivo functions of WRN helicase and exonuclease.

Acknowledgements

Research in the author's laboratory is supported by grants from the U.S.-Israel Binational Science Foundation, The Israel Science Foundation, the Conquer Fragile X Foundation Inc., The Chief Scientist—Israel Ministry of Health, and the Fund for Promotion of Research in the Technion.

References

1. Goto M, Rubenstein M, Weber J et al. Genetic linkage of Werner's syndrome to five markers on chromosome 8. Nature 1992; 355:735-738.
2. Schellenberg GD, Martin GM, Wijsman EM et al. Homozygosity mapping and Werner's syndrome. Lancet 1992; 339:1002.
3. Thomas W, Rubenstein M, Goto M et al. A genetic analysis of the Werner syndrome region on human chromosome 8p. Genomics 1993; 16:685-690.
4. Nakura J, Wijsman EJ, Miki T et al. Homozygosity mapping of the Werner syndrome locus (WRN). Genomics 1994; 23:600-608.
5. Oshima J, Yu CE, Boehnke M et al. Integrated mapping analysis of the Werner syndrome region of chromosome 8. Genomics 1994; 23:100-113.
6. Kihara K, Nakura J, Ye L et al. Carrier detection of Werner's syndrome using a microsatellite that exhibits linkage disequilibrium with the Werner's syndrome locus Jpn. J Hum Genet 1994; 39:403-409.
7. Ye L, Nakura J, Mitsuda N et al. Genetic association between chromosome 8 microsatellite (MS8-134) and Werner syndrome (WRN): Chromosome microdissection and homozygosity mapping. Genomics 1995; 28:566-569.
8. Yu C-E, Oshima J, Fu Y-H et al. Positional cloning of the Werner's syndrome gene. Science 1996; 272:258-262.
9. Moser MJ, Oshima J, Monnat RJ Jr. WRN mutations in Werner syndrome. Hum Mutat 1999; 13:271-279.
10. Gray MD, Shen JC, Kamath-Loeb AS et al. The Werner syndrome protein is a DNA helicase. Nat Genet 1997; 17:100-103.
11. Suzuki N, Shimamoto A, Imamura O et al. DNA helicase activity in Werner's syndrome gene product synthesized in a baculovirus system. Nucl Acids Res 1997; 25:2973-2978.
12. Ye L, Nakura J, Morishima A et al. Transcriptional activation by the Werner syndrome gene product in yeast. Exp Gerontol 1998; 33:805-812.
13. Balajee AS, Machwe A, May A et al. The Werner syndrome protein is involved in RNA polymerase II transcription. Mol Biol Cell 1999; 10:2655-2668.
14. Mushegian AR, Bassett DE Jr, Boguski MS et al. Positionally cloned human disease genes: Patterns of evolutionary conservation and functional motifs Proc Natl Acad Sci USA 1997; 94:5831-5836
15. Morozov V, Mushegian AR, Koonin EV et al. A putative nucleic acid-binding domain in Bloom's and Werner's syndrome helicases. Trends Biochem Sci 1997; 22:417-418.
16. Moser MJ, Holley WR, Chatterjee A et al. The proofreading domain of Escherichia coli DNA polymerase I and other DNA and/or RNA exonuclease domains. Nucl Acids Res 1997; 25: 5110-5118.
17. Shen JC, Gray MD, Oshima J et al. Werner syndrome protein. I. DNA helicase and DNA exonuclease reside on the same polypeptide. J Biol Chem 1998; 273:34139-34144.
18. Kamath-Loeb AS, Shen JC, Loeb LA et al. Werner syndrome protein. II. Characterization of the integral 3' → 5' DNA exonuclease. J Biol Chem 1998; 273:34145-34150.
19. Huang S, Li B, Gray MD et al. The premature ageing syndrome protein, WRN, is a 3'→5' exonuclease. Nat Genet 1998; 20:114-116.
20. Matsumoto A, Imamura O, Goto M et al. Characterization of the nuclear localization signal in the DNA helicase involved in Werner's syndrome. Intl J Mol Med 1998; 1:71-76.
21. Matsumoto T, Shimamoto A, Goto M et al. Impaired nuclear localization of defective DNA helicases in Werner's syndrome. Nat Genet 1997; 16:335-336.
22. Wang L, Hunt KE, Martin GM et al. Structure and function of the Werner syndrome gene promoter: evidence for transcriptional modulation. Nucl Acids Res 1998; 26:3480-3485.
23. Umezu K, Nakayama K, Nakayama H. Escherichia coli RecQ protein is a DNA helicase Proc Natl Acad Sci USA 1990; 87:5363-5367.
24. Watt PM, Louis EJ, Borts R et al. Sgs1: A eukaryotic homolog of E. coli RecQ that interacts with DNA topoisomerase II in vivo and is required for faithful chromosome segregation. Cell 1995; 81:253-260.

25. Stewart E, Chapman CR, Al-Khodairy F et al. rqh⁺, a fission yeast gene related to the Bloom's and Werner's syndrome genes, is required for reversible S phase arrest. EMBO J 1997; 16:2682-2692.
26. Yan H, Chen CY, Kobayashi R et al. Replication focus-forming activity 1 and the Werner syndrome gene product. Nat Genet 1998; 19:375-378.
27. Imamura O, Ichikawa K, Yamabe Y et al. Cloning of a mouse homologue of the human Werner syndrome gene and assignment to 8A4 by fluorescence in situ hybridization. Genomics 1997; 41, 298-300.
28. Ellis NA, Groden J, Ye T–Z et al. The Bloom's syndrome gene product is homologous to RecQ helicases. Cell 1995; 83:655-666.
29. Seki M, Miyazawa H, Tada S et al. Molecular cloning of cDNA encoding human DNA helicase Q1 which has homology to Escherichia coli Rec Q helicase and localization of the gene at chromosome 12p12. Nucl Acids Res 1994; 22:4566-4573.
30. Puranam KL, Blackshear PJ. Cloning and characterization of RECQL, a potential human homologue of the Escherichia coli DNA helicase. RecQ J Biol Chem 1994; 269:29838-29845.
31. Kitao S, Ohsugi I, Ichikawa K et al. Cloning of two new human helicase genes of the RecQ family: biological significance of multiple species in higher eukaryotes. Genomics 1998; 54:443-452.
32. Kitao S, Lindor NM, Shiratori M et al. Rothmund-Thomson syndrome responsible gene, RECQL4: Genomic structure and products. Genomics 1999; 61:268-276.
33. Sekelsky JJ, Brodsky MH, Rubin GM et al. Drosophila and human RecQ5 exist in different isoforms generated by alternative splicing. Nucl Acids Res 1999; 27:3762-3769.
34. Bahr A, Graeve F, Kedinger C et al. Point mutations causing Bloom's syndrome abolish ATPase and DNA helicase activities of the BLM protein. Oncogene 1998; 17:2565-2571.
35. Kitao S, Shimamoto A, Goto M et al. Mutations in RECQL4 cause a subset of cases of Rothmund-Thomson syndrome. Nat genet 1999; 22:82-84.
36. Lindor NM, Furuichi Y, Kitao S et al. Rothmund-Thomson syndrome due to RECQ4 helicase mutations: report and clinical and molecular comparisons with Bloom syndrome and Werner syndrome. Am J Med Genet 2000; 90:223-228.
37. Mohaghegh P, Hickson ID. DNA helicase deficiencies associated with cancer predisposition and premature ageing disorders. Hum Mol Genet 2001; 10:741-746.
38. Shen J-C, Gray MD, Oshima J et al. Characterization of Werner syndrome DNA helicase activity: Directionality, substrate dependence and stimulation by replication protein A. Nucl Acids Res 1998; 26:2879-2885.
39. Umezu K, Nakayama K, Nakayama H. Escherichia coli RecQ protein is a DNA helicase Proc Natl Acad Sci USA 1990; 87:5363-5367
40. Henderson E, Hardin CC, Walk SK et al. Telomeric DNA oligonucleotides form novel intramolecular structures containing guanine-guanine base pairs. Cell 1987; 51:899-908.
41. Sen D, Gilbert W. Formation of parallel four-stranded complexes by guanine-rich motifs in DNA and its implications for meiosis. Nature 1988; 334:364-366.
42. Sundquist WI, Klug A. Telomeric DNA dimerizes by formation of guanine tetrads between hairpin loops. Nature 1989; 342:825-829.
43. Sen D, Gilbert W. A sodium-potassium switch in the formation of four-stranded G4-DNA. Nature 1990; 344:410-414.
44. Zahler AM, Williamson JR, Cech TR et al. Inhibition of telomerase by G-quartet DNA structures Nature 1991; 350:718-720.
45. Williamson JR. G-quartet structures in telomeric DNA. Annu Rev Biophys Biomol Struct 1994; 23:703-730.
46. Kamath-Loeb AS, Loeb LA, Johansson E et al. Interaction between Werner syndrome helicase and polymerase delta specifically facilitate copying of tetraplex and hairpin structures of the $d(CGG)_n$ trinucleotide repeat sequence. J Biol Chem 2001; 276:16439-16446.
47. Fry M, Loeb LA. The fragile X syndrome $d(CGG)_n$ nucleotide repeats form a stable tetrahelical structure. Proc Natl Acad Sci 1994; 91:4950-4954.
48. Fry M, Loeb LA. Human Werner syndrome DNA helicase unwinds tetrahelical structures of the fragile X repeat sequence $d(CGG)_n$. J Biol Chem 1999; 274:12797-12802.
49. Mohaghegh P, Karow JK, Brosh RM Jr et al. The Bloom's and Werner's syndrome proteins are DNA structure-specific helicases. Nucl Acids Res 2001; 29:2843-2849.

50. Sun H, Karow JK, Hickson ID et al. The Bloom's syndrome helicase unwinds G4 DNA. J Biol Chem 1998; 273:27587-27592.
51. Sun H, Bennet RJ, Maizels N. The Saccharomyces cerevisiae Sgs1 helicase efficiently unwinds G-G paired DNAs. Nucl Acids Res 1999; 27:1978-1984.
52. Kamath-Loeb AS, Johansson E, Burgers PMJ et al. Functional interaction between Werner syndrome protein and DNA polymerase d. Proc Natl Acad Sci USA 2000; 97:4603-4608.
53. Szekely AM, Chen Y–H, Zhang C et al. Werner protein recruits DNA polymerase d to the nucleolus. Proc Natl Acad Sci USA 2000; 97:11365-11370.
54. Schultz VP, Zakian VA, Ogburn CE et al. Accelerated loss of telomeric repeats may not explain accelerated replicative decline of Werner syndrome cells. Hum Genet 1996; 97:750-754.
55. Tahara H, Tokutake Y, Maeda S et al. Abnormal telomere dynamics of B-lymphoblastoid cell strains from Werner's syndrome patients transformed by Epstein-Barr virus. Oncogene 1997; 15:1911-1920.
56. Wyllie FS, Jones CJ, Skinner JW et al. Telomerase prevents the accelerated cell ageing of Werner syndrome fibroblasts. Nat Genet 2000; 24:16-17.
57. Ouellette MM, McDaniel LD, Wright WE et al. The establishment of telomerase-immortalized cell lines representing human chromosome instability syndromes. Hum Mol Genet 2000; 9:403-411.
58. Choi D, Whittier PS, Oshima J et al. Telomerase expression prevents replicative senescence but does not fully reset mRNA expression patterns in Werner syndrome cell strains. FASEB J 2001; 15:1014-1020.
59. Johnson FB, Marciniak RA, McVey M et al. The Saccharomyces cerevisiae homolog Sgs1p participates in telomere maintenance in cells lacking telomeres. EMBO J 2001; 20:905-913.
60. Li J-L, Harrison J, Reszka AP et al. Inhibition of the Bloom's and Werner's syndrome helicases by G-quadruplex interacting ligands. Biochemistry 2001; 40:15194-15202.
61. Prince PR, Emond MJ, Monnat RJ. Loss of Werner syndrome protein function promotes aberrant mitotic recombination. Genes Dev 2001; 15:933-938.
62. Karow JK, Constantinou A, Li J-L et al. The Bloom's syndrome gene product promotes branch migration of Holliday junctions. Proc Natl Acad Sci USA 2000; 97:6504-6508.
63. Constantinou A, Tarsounas M, Karow JK et al. Werner's syndrome protein (WRN) migrates Holliday junctions and co-localizes with RPA upon replication arrest. EMBO J 2001; Rep 1: 80-84.
64. Frank-Kamenetskii MD, Mirkin SM. Triplex DNA structures. Annu Rev Biochem 1995; 64:65-95.
65. Agazie YM, Burkholder GD, Lee JS. Triplex DNA in the nucleus: Direct binding of triplex-specific antibodies and their effect on transcription, replication and cell growth. Biochem J 1996; 316:461-466.
66. Guieysse AL, Praseuth D, Helene C. Identification of a triplex DNA-binding protein from human cells. J Mol Biol 1997; 267:289-298.
67. Musso M, Nelson LD, Van Dyke MW. Characterization of purine-motif triplex DNA-binding proteins in HeLa extracts. Biochemistry 1998; 37:3086-3095.
68. Nelson LD, Musso M, Van Dyke MW. The yeast STM1 gene encodes a purine motif triple helical DNA-binding protein. J Biol Chem 2000; 275:5573-5581.
69. Brosh RM Jr, Majumdar A, Desai S et al. Unwinding of a DNA triple helix by the Werner and Bloom syndrome helicases. J Biol Chem 2001; 276:3024-3030.
70. Kohwi Y, Panchenko Y. Transcription-dependent recombination induced by triple-helix formation. Genes Dev 1993; 7:1766-1778.
71. Rooney SM, Moore PD. Antiparallel, intramolecular triplex DNA stimulates homologous recombination in human cells. Proc Natl Acad Sci. USA 1995; 92:2141-2144.
72. Benet A, Azorin F. The formation of triple-stranded DNA prevents spontaneous branch-migration. J Mol Biol 1999; 294:851-857.
73. Harmon FG, Kowalczykowski SC. RecQ helicase, in concert with RecA and SSB proteins, initiates and disrupts DNA recombination. Genes Dev 1998; 12:1134-1144.
74. Shen J-C, Loeb LA. Werner syndrome exonuclease catalyzes structure-dependent degradation of DNA. Nucl Acids Res 2000; 28:3260-3268.
75. Machwe A, Xiao L, Theodore S et al. DNase I footprinting and enhanced exonuclease function of a bipartite Werner syndrome protein (WRN) bound to partially melted duplex DNA. J Biol Chem 2002; 277:4492-4504.

76. Orren DK, Brosh RM Jr, Nehlin JO et al. Enzymatic and DNA binding properties of purified WRN protein: high affinity binding to single-stranded DNA but not to DNA damage induced by 4NQO. Nucl Acids Res 1999; 27:3557-3566.
77. Gebhart E, Bauer R, Raub U et al. Spontaneous and induced chromosomal instability in Werner syndrome. Hum Genet 1988; 80:135-139.
78. Ogburn CE, Oshima J, Poot M et al. An apoptosis-inducing genotoxin differentiates heterozygotic carriers for Werner helicase mutations from wild-type and homozygous mutants. Hum Genet 1997; 101:121-125.
79. Brosh RM Jr, Karow JK, White EJ et al. Potent inhibition of Werner and Bloom helicases by DNA minor groove binding drugs. Nucl Acids Res 2000; 28:2420-2430.
80. Duan W, Rangan A, Vankayalapati H et al. Design and synthesis of fluoroquinophenoxazines that interact with human telomeric G-quadruplexes and their biological effects. Mol Cancer Therap 2001; 1:103-120.
81. Shi DF, Wheelhouse RT, Sun D et al. Quadruplex-interactive agents as telomerase inhibitors: synthesis of porphyrins and structure-activity relationship for the inhibition of telomerase. J Med Chem 2001; 44:4509-4523.
82. Kim MY, Vankayalapati H, Shin-Ya K et al. Telomestatin, a potent telomerase inhibitor that interacts quite specifically with the human telomeric intramolecular G-quadruplex. J Am Chem Soc 2002; 124:2098-2099.
83. Riou JF, Guittat L, Mailliet P et al. Cell senescence and telomere shortening induced by a new series of specific G-quadruplex DNA ligands. Proc Natl Acad Sci USA 2002; 99:2672-2677.
84. Suzuki N, Shiratori M, Goto M et al. Werner syndrome helicase contains a 5'→3' exonuclease activity that digests DNA and RNA strands in DNA/DNA and RNA/DNA duplexes dependent on unwinding. Nucl Acids Res 1999; 27:2361-2368.
85. Cooper MP, Machwe A, Orren DK et al. Ku complex interacts with and stimulates the Werner protein. Genes Dev 2000; 14:907-912
86. Li B, Comai L. Functional interaction between Ku and the Werner syndrome protein in DNA end processing. J Biol Chem 2000; 275:28349-28352.
87. Huang S, Beresten S, Li B et al. Characterization of the human and mouse WRN 3'→5' exonuclease. Nucleic Acids Res 2000; 28:2396-2405.
88. Li B, Comai L. Requirements for the nucleolytic processing of DNA ends by the Werner syndrome protein-Ku 70/80 complex. J Biol Chem 2001; 276:9896-9902.
89. Machwe A, Xiao L, Theodore S et al. DNase I footprinting and enhanced exonuclease function of the bipartite Werner syndrome protein (WRN) bound to partially melted duplex DNA. J Biol Chem 2002; 277:4492-4504.
90. Xue Y, Ratcliff GC, Wang H et al. A minimal exonuclease domain of WRN forms a hexamer on DNA and possesses both 3'→5' exonuclease and 5'-protruding strand endonuclease activities. Biochemistry 2002; 41:2901-2912.
91. Brosh RM Jr, Karmakar P, Sommers JA et al. p53 modulates the exonuclease activity of Werner syndrome protein. J Biol Chem 2001; 276:35093-35102.
92. Blander G, Kipnis J, Leal JFM et al. Physical and functional interaction between p53 and the Werner syndrome protein. J Biol Chem 1999; 275:29463-29469.
93. Karow JK, Newman RH, Freemont PS et al. Oligomeric ring structure of the Bloom's syndrome helicase. Curr Biol 1999; 9:597-600.
94. Lohman TM, Bjornson KP. Mechanisms of helicase-catalyzed DNA unwinding. Annu Rev Biochem 1996; 65:169-214.
95. Opresko PL, Laine J–P, Brosh RM Jr et al. Coordinate action of the helicase and 3' to 5' exonuclease of Werner syndrome protein. J Biol Chem 2001; 276:44677-44687.
96. Gray MD, Wang L, Youssoufian H et al. Werner helicase is localized to transcriptionally active nucleoli in cycling cells. Exp Cell Res 1998; 242:487-494.
97. Marciniak RA, Lombard DB, Johnson FB et al. Nucleolar localization of the Werner syndrome protein in human cells. Proc Natl Acad Sci USA 1998; 95:6887-6892.
98. Suzuki T, Shiratori M, Furuichi Y et al. Diverged nuclear localization of Werner helicase in human and mouse cells. Oncogene 2001; 20:2551-2558.

99. Machwe A, Orren DK, Bohr VA. Accelerated methylation of ribosomal RNA genes during the cellular senescence of Werner syndrome fibroblasts. FASEB J 2000; 14:1715-1724.
100. Fry M. The Werner Syndrome Helicase-Nuclease – One Protein, Many Mysteries Science. SAGE KE 2002; http://sageke.sciencemag.org/cgi/content/full/sageke;2002/13/re2 10pp.
101. Shen J-C, Loeb LA. Unwinding the molecular basis of the Werner syndrome. Mech Ageing Dev 2001; 122:921-944.
102. Brosh RM Jr, Waheed J, Sommers JA. Biochemical characterization of the DNA substrate specificity of Werner syndrome helicase. J Biol Chem 2002; 277:23236-23245.
103. Yang Q, Zhang R, Wang XW et al. J Biol Chem 2002; 277:31980-31987.
104. Orren DK, Theodore S, Machwe A. The Werner syndrome helicase/exonuclease (WRN) disrupts and degrades D-loops in vitro. Biochemistry 2002; 41:13483-13488.
105. Karmakar P, Piotrowski J, Brosh RM Jr et al. Werner protein is a target of DNA-dependent protein kinase in vivo and in vitro, and its catalytic activities are regulated by phosphorylation. J Biol Chem 2002; 277:18291-18302.

CHAPTER 4

Proteins That Interact with the Werner Syndrome Gene Product

Dana Branzei and Takemi Enomoto

Abstract

Werner syndrome is an autosomal recessive disorder characterized by premature onset of age-related diseases, increased cancer incidence, and genomic instability. Biochemical characterization has shown WRN protein to have helicase and exonuclease activities of 3'-5' polarity, and to have an associated ATPase activity. However, the molecular mechanism of WRN-DNA transactions is not yet fully understood. Elucidation of the physiologic functions of WRN may be aided by the identification of WRN-interacting proteins, and the processes into which these proteins are involved. In this chapter, we review the proteins that were found to interact with WRN, and in the light of our current view of the roles of WRN in cells and the functions of these WRN-interacting proteins, we discuss how these interactions might affect WRN functions in DNA metabolic processes such as replication, recombination and repair to preserve the genomic integrity in cells.

Introduction

Although the biochemical studies of WRN have uncovered the catalytic activities of this protein and have suggested several DNA processes into which WRN might participate, it is not yet fully understood how mutations in this protein lead to a premature aging phenotype and to genomic instability. It was therefore thought that by identifying physiologically interacting proteins, the principal pathways into which WRN participates could be delineated. Furthermore, it is thought that mutations in genes encoding WRN-interacting proteins may be responsible for about 10% of patients who fulfill clinical and diagnostic criteria for WRN, but lack WRN mutations.[1] A series of WRN interacting proteins have been identified based on physical interactions and functional assays. They include proteins involved in DNA replication (RPA, PCNA, DNA Polymerase δ, Topoisomerase I and II, FEN-1 and EXO-1, WHIP/WRNIP1, PARP-1), checkpoint control (p53), homologous recombination and repair (RAD51, RAD52, RPA), and non-homologous end joining (NHEJ) processes (DNA-PK$_{cs}$, Ku). A list of the WRN-interacting proteins identified to date is shown in Table 1, and the main processes into which these proteins are implicated are illustrated in Figure 1. Here we introduce these proteins by order of their principal function and then discuss how these interactions may affect WRN functions in cells.

Werner Interacting Proteins

DNA Replication Proteins

Several lines of evidence suggest that WRN may be implicated in some aspects of DNA replication. The S phase of WS cells is prolonged[2-4] with a reduced frequency of replicon

---

Molecular Mechanisms of Werner's Syndrome, edited by Michel Lebel. ©2004 Eurekah.com and Kluwer Academic / Plenum Publishers.

Table 1. WRN interacting proteins

WRN Interacting Protein	Description
RPA	ssDNA binding protein, functions in replication, repair and recombination
PCNA	Replication clamp, functions in replication, repair, and recombination
Topoisomerase I	Functions to relax supercoiling and relieve torsional stress
Topoisomerase II	Functions in the decatenation process of chromosomes
DNA Polymerase δ	Major replicative DNA polymerase
FEN-1	5' flap endonuclease/ 5'-3' exonuclease, functions during Okazaki fragment processing
EXO-1	RAD2 family exonuclease and endonuclease, functions in base-excision repair
DNA Polymerase β	Functions in short-patch and long-patch base excision repair
WHIP/WRNIP1	Protein similar to replication factor C subunits, likely to function in DNA replication
p53	Tumor suppressor gene product, functions in checkpoint control and apoptosis
RAD52	Recombination protein, mediates single strand annealing, strand exchange, and Holliday junction formation
RAD51	Recombination protein, functions in strand exchange
BLM	RecQ helicase, functions in HJ resolution
Ku70-Ku80	DNA duplex end binding protein, functions in NHEJ
DNA-PK$_{cs}$	The catalytic subunit of the DNA-PK, functions in NHEJ
p97/VCP	Functions in protein folding and protein degradation
Ubc9, SUMO-1	SUMO-1 modification pathway

initiation,[5] and the homologue of Werner protein in *Xenopus laevis* is required for replication foci formation.[6] Furthermore, WRN translocates to sites of replication/repair when replication is blocked by hydroxyurea,[7] co-purifies with the DNA replication complex,[8,9] and binds several components of the DNA replication fork.

RPA

The human replication protein A (RPA) is a single-stranded DNA binding protein, which functions to stabilize the interaction between RFC, PCNA, and ssDNA during replication and to modulate the function of other proteins in processes of repair, replication, and recombination. RPA stimulates the helicase activity of WRN,[10,11] and is required by WRN in unwinding alternative DNA structures such as telomere repeat complexes.[12] Physical interaction between WRN and RPA was demonstrated by co-immunoprecipitation,[11] and co-localization studies have shown that WRN foci coincide with RPA foci at sites of stalled replication forks caused by depletion of cellular dNTP pool after hydroxyurea treatment.[7] A different study found that WRN forms nuclear foci in response to various DNA damaging agents, such as camptothecin, etoposide, 4-nitroquinoline-N-oxide and bleomycin, and that these WRN foci overlap almost completely with the foci of RPA.[13] These findings suggest that WRN and RPA function cooperatively in processes such as double-strand break repair and processing of stalled replication forks.

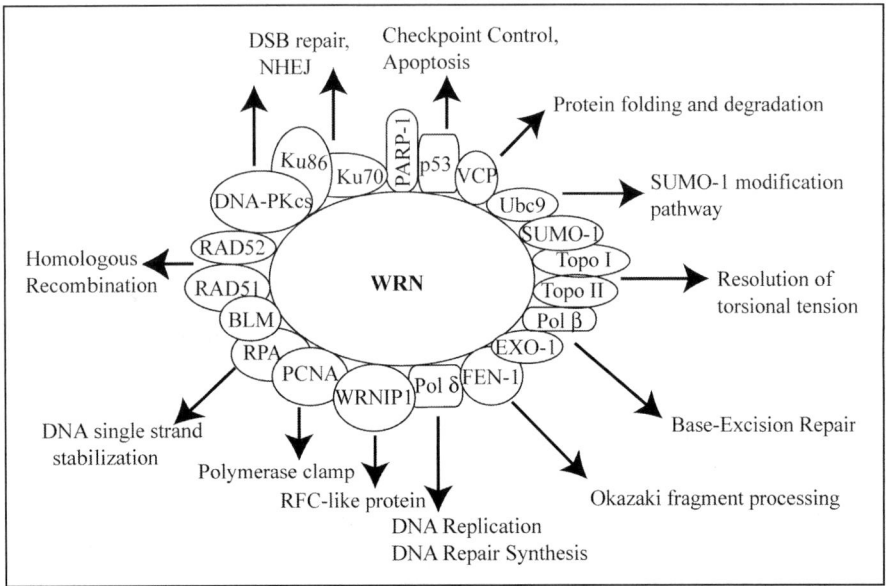

Figure 1. Proteins that interact with WRN and their proposed functions. WRN was found to interact with proteins involved in replication (RPA, PCNA, Polymerase δ, Topoisomerase I, Topoisomerase II, FEN-1, WHIP/WRNIP), BASE excision repair (EXO-1/DNA polymerase β), checkpoint control (p53), homologous recombination (RPA, RAD51, RAD52, BLM), various processes of DNA metabolism (PARP-1), non-homologous end joining processes (DNA-PK$_{cs}$, Ku), protein degradation (p97/VCP), and SUMO-1 modification (UBC9/ SUMO-1).

PCNA, Topoisomerase I, and Topoisomerase II

It was found that the wild type WS protein co-purifies with the 17S multiprotein DNA replication complex, while the mutant WS protein fails to associate with the complex, indicating that this interaction is a physiological one.[9] In addition, WRN interacts directly with PCNA, topoisomerase I, and RPA. PCNA was found to bind to the N-terminal of the Werner protein, which contains the exonuclease domain (amino acids 168 to 246). The interaction between WRN and PCNA is interesting, because PCNA is a potential communication point between a variety of important cellular processes, including cell cycle control, DNA replication, DNA repair, and recombination.[14]

In addition, topoisomerase II co-immunoprecipitates with WRN from cells treated with the catalytic inhibitor ICRF187.[15] IRF187 inhibits the ATPase activity of topoisomerase II and stabilizes the enzyme in the form of a closed clamp without inducing formation of DNA breaks. Topoisomerase II is required for the decatenation process of chromosomes (or the separation of newly synthesized chromosomes) immediately after DNA replication, and inhibition of its activity results in activation of the decatenation checkpoint. WS cells were found to be defective in the decatenation checkpoint. Furthermore, override of the decatenation checkpoint results in increased apoptosis and DNA damage only in the absence of WRN, but not in normal cells.[15] These results implicate an important role for WRN in the handling and repair of DNA strand breaks resulting from replicating incomplete catenated chromatids.

DNA Polymerase δ

DNA polymerase δ is a major replicative DNA polymerase in eukaryotic cells that together with polymerase α and ε accounts for genomic replication. The discovery that WRN interacts physically and functionally with polymerase δ brought further support to the idea that WRN is involved in DNA replication. By using the two-hybrid screening, it was found that the C-terminus of WRN interacts with the p50 subunit of the human polymerase δ.[16] Native WRN co-immunoprecipitates with p50 in a cellular fraction enriched in nucleolar proteins, and p125, the catalytic subunit of DNA polymerase δ is also present in this nucleolar fraction. Furthermore, subcellular localization studies of cells transfected with WRN found that p50 and p125 redistribute to the nucleolus and co-localize with WRN. These results suggest that WRN could be involved in regulating DNA replication by recruiting polymerase δ to particular sites of DNA synthesis.

It has been reported that WRN enhances the rate of nucleotide incorporation by yeast polymerase δ in the absence of PCNA.[17] This functional interaction is specific to polymerase δ, and is mediated by the third subunit of polymerase δ, Pol32 of *S. cerevisiae*, which corresponds to the p66 subunit of human polymerase δ. In contrast, it was found that WRN has no significant stimulatory effect on the DNA polymerase δ holoenzyme (polymerase δ-PCNA complex),[17] and as such, it is unlikely that WRN functions in the DNA replication process per se. Rather, these results suggest that WRN may function in a replication restart pathway at sites where unusual DNA structures have blocked DNA replication or where the DNA replication machinery has detached from the DNA. Consistent with this view, WRN was shown to relieve the constraints imposed by template secondary structures on polymerase δ-catalyzed synthesis.[18] Again, this functional interaction seemed to be specific to polymerase δ among DNA polymerases, and to RecQ among helicases.[18]

Taken together, these findings suggest that WRN might facilitate polymerase δ mediated DNA transactions, such as DNA replication and repair, and a defective WRN-polymerase δ interaction in WS cells might contribute to the S-phase defects of these cells.

DNA Polymerase β

DNA polymerase β is an important enzyme involved in cellular short-patch and long-patch base excision repair mechanisms. WRN stimulates strand displacement synthesis activity of DNA polymerase β in vitro, and co-immunoprecipitation experiments have indicated that WRN and DNA polymerase β are part of the same complex in the cell.[19] Thus, this interaction suggests the involvement of WRN in such DNA repair mechanisms.

FEN-1 and EXO-1

WRN was also reported to interact physically with FEN-1 through a direct protein interaction.[20] FEN-1 is a 5' flap endonuclease/ 5'-3' exonuclease required during Okazaki fragment processing and long patch base-excision repair. The FEN-1 cleavage activity is strongly stimulated by WRN, in a process that does not require the catalytic activities (ATPase, helicase, or exonuclease activities) of WRN. Interestingly, neither human PCNA nor human RPA, which were reported to interact and stimulate FEN-1 cleavage activity, displayed a stimulation of FEN-1 cleavage activities that approached the stimulation by WRN. The functional interaction with FEN-1 is mediated through a C-terminal domain of WRN containing the 949-1042 amino acid region. Among RecQ helicases, this region of WRN shows the highest degree of homology with FFA-1, for which a role in replication has been implicated.[21,22]

The human exonuclease-1 (EXO-1) is another member of the RAD2 family to which FEN-1 belongs.[23] In fact, functional overlaps between EXO-1 and FEN-1 have been proposed from yeast work.[24] WRN was shown to form a complex with EXO-1 in vivo and to stimulate the exonucleolytic and endonucleolytic activities of EXO-1 in vitro.[25] Similar to the FEN-1 report,[20] the C-terminal region of WRN was found to be necessary for this stimulation, while mutations in the exonuclease domain or the helicase domain of WRN did not affect its stimulatory effect on EXO-1 activities. These facts suggest that WRN may function with FEN-1 and EXO-1 during replication to ensure efficient and accurate processing of specific DNA structures that arise during DNA synthesis.

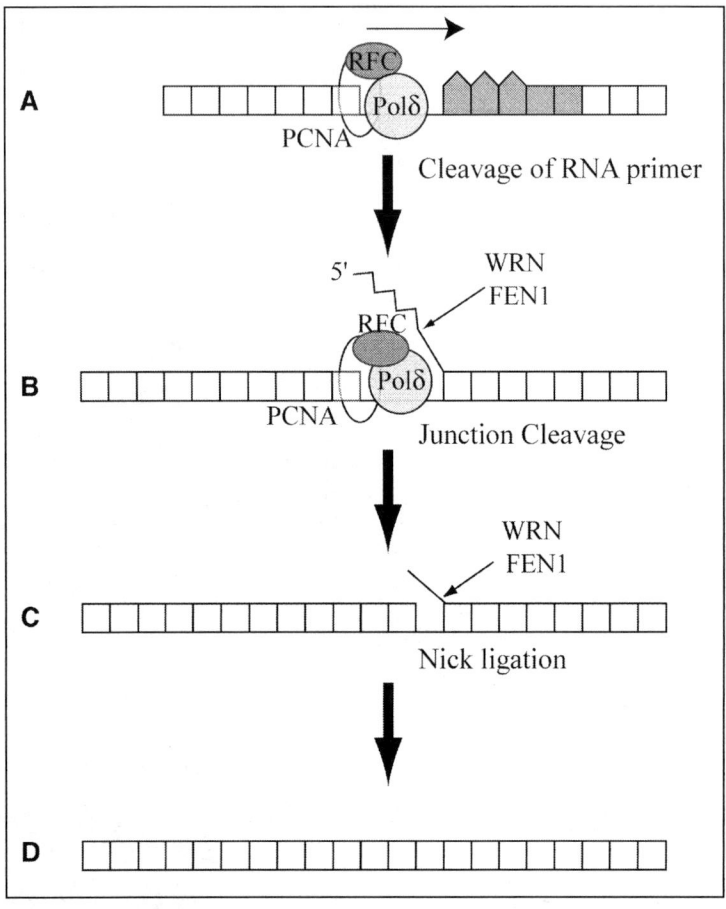

Figure 2. Possible role for WRN-FEN1 in Okazaki fragment processing. A) Displacement synthesis by Pol δ results in the formation of an RNA-DNA flap structure. B) WRN might stimulate FEN-1 to cleave the RNA portion of the flap-structure. C) Alternatively, WRN might promote FEN-1-mediated cleavage of 5'-flap structures formed during displacement synthesis on the lagging strand or the junction of the Okazaki fragment. D) Final ligation of the nick completes the process.

WRN might function to stimulate FEN-1 activity during lagging strand synthesis at several steps. For instance, WRN might modulate the FEN-1-mediated incision of the remaining ribonucleotide in conjunction with displacement synthesis of Okazaki fragments by Pol δ, after which ligation of Okazaki fragments can take place (Fig. 2). Since dissociation of polymerase δ is necessary in order to allow FEN-1 access to junction cleavage, it is possible that WRN promotes both dissociation of the Pol δ complex (RFC, PCNA, Pol δ) from DNA and the FEN-1 mediated cleavage of the junction (Fig. 2B). Alternatively, WRN might stimulate the FEN-1 mediated cleavage of longer 5' flap structures that arise during strand displacement synthesis on the lagging strand, or simply to unwind the RNA (DNA) primer of an Okazaki fragment during Okazaki fragment processing.

Another possibility is that, during replication restart, WRN might translocate to a stalled replication fork, and promote removal of the nascent Okazaki fragment, by displacing it and stimulating the nuclease activities of FEN-1 (see Fig. 3), in a model similar to that proposed by Courcelle and Hanawalt for RecQ and RecJ.[26] This could also take place when abnormal

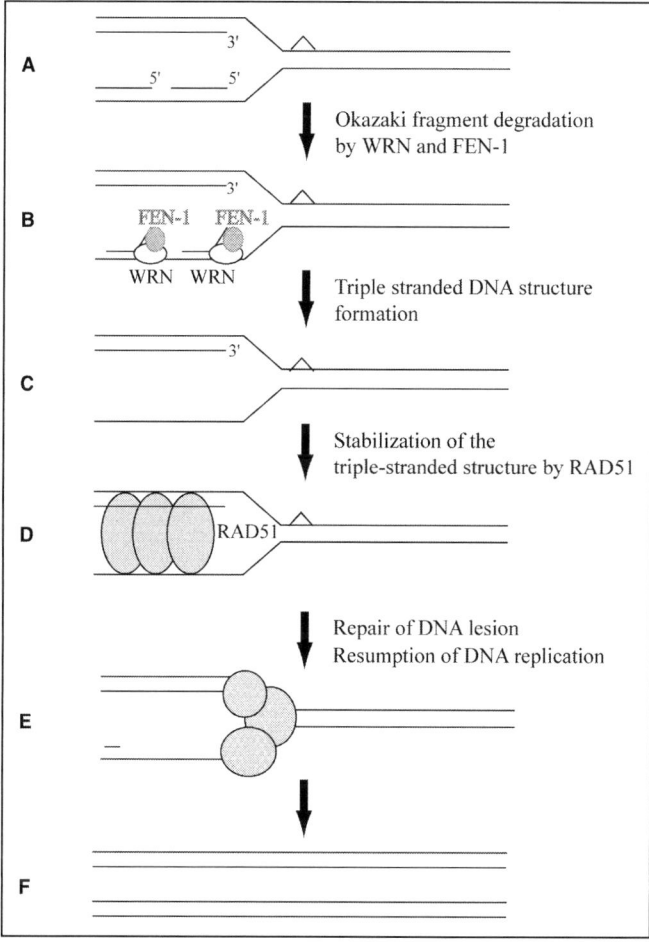

Figure 3. Model for WRN function at a stalled replication fork. A) Replication fork encounters replication block and stalls. B) By analogy to RecQ and RecJ, WRN and FEN-1 might act to displace and cleave Okazaki fragments, C) giving rise to a triple-stranded structure. D) Rad51 and WRN may act to stabilize the triple-stranded structure, E) until the replication block is removed and replication can resume. F) Replication can thus be completed without strand breakage and recombination. This model was originally proposed for RecQ and RecJ by Courcelle and Hanawalt.[26]

Okazaki fragments have been synthesized or when lagging strand synthesis has been blocked by unusual DNA structures. This mechanism would prevent deleterious strand breakage or recombination and would help maintain genomic stability.

WHIP/WRNIP1

WHIP/WRNIP1 is a novel protein identified as a Werner interacting protein (WRN interacting protein) in a two-hybrid screening.[27] Like RPA, PCNA, and topoisomerase I, WRNIP1 binds to the N-terminus of WRN (amino acids 1 to 272), which contains the exonuclease activity. WRNIP1 interaction seems to be specific to WRN among RecQ helicases, since no interaction was detected by the yeast two-hybrid system between WRNIP1 and BLM, or WRNIP1 and RECQL1. Ectopically expressed WRNIP1 and WRN were shown to co-localize in granular structures in the nucleus.[27]

WRNIP1 is a highly conserved protein, homologues of which are widely distributed in eubacteria and eukaryotes. The eukaryote WRNIP1s show a high percent of identity and similarity to replication factor C subunits and some homology to RuvB.[28] Furthermore, the RFC motifs found in *S. cerevisiae* Wrnip1/Mgs1 are essential for Mgs1 function in DNA recombination, mutagenesis, and interaction with other replication proteins, corroborating to the idea that Mgs1 plays a role in DNA replication.[29] The bacterial homologue, designated YcaJ or RarA[30] is closely related to both the Holliday junction helicase, RuvB, and to DnaX, which encodes the τ and γ components of the *E. coli* DNA polymerase III replisome. Both bacterial and budding yeast WRNIP1 homologues display DNA dependent ATPase activity, and DNA annealing activity,[28] and seem to play a role in influencing recombination outcome.[28,30,31] In *S. cerevisiae*, deletion of the *WRNIP1/MGS1* gene in combination with *top3* and *sgs1* results in slow-growth and increased genomic instability,[28,31] and leads to lethality in combination with mutations in the post-replicating repair genes *rad6* or *rad18*,[32] suggesting a role for this gene in post-replication repair and in maintaining genomic integrity by controlling mitotic recombination.

Although a functional interaction between WRN and WRNIP1 has not yet been observed, studies in *S. cerevisiae* showed that simultaneous deletion of *MGS1* and *SGS1* (a yeast orthologue of *WRN*) leads to a slow-growth phenotype accompanied by reduced life-span, poor cell viability and a synergistic increase in recombination frequencies,[27,31,32] indicating that the function of *MGS1* is important for maintaining genomic stability in the absence of a functional *SGS1* gene. In addition, deletion of *MGS1* gene partially alleviates the sensitivity to methyl methanesulfonate of *sgs1* disruptants, suggesting that under DNA damage conditions, Mgs1 might promote an Sgs1-dependent repair pathway.[27]

Finally, in *S. cerevisiae*, genetic and functional interactions have been observed between *MGS1* and genes encoding subunits of Polymerase δ, PCNA, and RFC subunits, which suggest that Mgs1 interacts with the DNA replication machinery to modulate the function of DNA polymerase δ during replication or replication-associated repair.[29,32] We speculate that WRNIP1 might provide a link between WRN and the replication machinery during DNA replication, and might be involved in regulating WRN functions during replication restart at stalled replication forks or during replication-associated repair by influencing the choice of the pathway employed for replication fork reactivation.

Poly(ADP-Ribosyl) Polymerase-1 (PARP-1)

Poly(ADP-ribose) polymerase-1 (PARP-1) is involved in nuclear processes involving cleavage and/or rejoining of DNA, such as DNA replication, DNA repair, recombination, and apoptosis. Inhibition of the PARP-1 results in an increased frequency of DNA strand breaks, recombination, gene amplification, micronuclei formation, and sister chromatid exchanges, in cells exposed to DNA damaging agents.[33] PARP-1 is also associated with the DNA replication apparatus and is considered a nick sensor during replication. It can also bind to double strand breaks. Finally, it optimizes base excision repair in cells.[34] Recent co-immunoprecipitation studies have demonstrated an interaction between WRN and PARP-1 in vivo.[35] Furthermore, crosses between Wrn mutant and PARP-1 null mice have shown a genetic cooperation between Wrn and PARP-1 in preventing chromatid breaks, complex chromosomal rearrangements, and cancer in these animals.[35] These results suggest that Wrn and PARP-1 may be part of a complex involved in the processing of DNA breaks. As PARP-1 and PARP-2 are known to form a heterodimer,[36] PARP-2 may also be part of such a complex.

Checkpoint Control Proteins-p53

The tumor suppressor gene product p53 plays a critical role in maintaining the genomic integrity of mammalian cells, perhaps through its ability to induce apoptosis. Direct interaction between p53 and WRN was demonstrated by co-immunoprecipitation and in vitro studies.[37,38] WRN protein binds to the carboxyl terminus of p53,[38] which is also the binding site of the DExH family-related DNA helicases XPB and XPD.[39] Moreover, fibroblasts from WS individuals show an attenuated ability to undergo p53-mediated apoptosis as compared with normal fibroblasts, and this deficiency can be restored by expression of WRN.[38] This observation suggests that WRN may be a downstream target in the p53-mediated apoptosis.

In cells arrested in S phase, WRN and p53 co-localize in the nucleoplasm. In addition, p53 inhibits the exonuclease activity of the purified full-length WRN protein, and for this inhibition the physical interaction domain of WRN with p53, which lies in the carboxyl terminus of WRN, is necessary.[40] Interestingly, two naturally occurring p53 mutants found in human cancer displayed a reduced ability to inhibit WRN exonuclease activity,[40] suggesting that modulation of WRN exonuclease activity by p53 might be important in preventing genomic instability.

In addition to its checkpoint function, p53 has been implicated in homologous recombination (HR), based on the facts that (A) p53 mutant cell lines show increased rates of HR, (B) p53 can bind and inhibit human RAD51 and bacterial RecA, which are pivotal components of the HR pathway, (C) p53 can bind to several DNA structures associated with HR, and (D) overexpression of p53 down-regulates the rate of HR between viral SV40 DNA molecules. WRN was shown to recognize and disrupt Holliday Junctions (HJ, an intermediate DNA molecule during HR), which arise during replication. It is believed that by doing so WRN functions to reduce inappropriate DNA recombination in vivo.[7] Recently, it was shown that purified recombinant p53 binds WRN and attenuates its ability to unwind synthetic Holliday Junctions in vitro.[41] However, this property can be decreased by phosphorylation of p53 C terminus at Ser^{376} or Ser^{378}. These results suggest a possible physiological mechanism for the regulation of HR through the physical and functional interaction of p53 with WRN (see Fig.4).

In addition, *p53 null/Wrn* helicase mutant mice rapidly and simultaneously develop multiple types of tumors, indicating that *Wrn* mutation induces accelerated tumorigenesis on a *p53* null background.[42]

Although the cellular levels of p53 in WS fibroblasts are not significantly different from normal cells,[38,43] overexpression of WRN in normal fibroblasts increases p53 levels and stimulate p53-mediated apoptosis.[44] On the other hand, overexpression of p53 down-regulates the promoter activity of WRN.[45] These results suggest the existence of a feedback control between p53 and WRN, and indicate that cooperation between these two proteins could be crucial in regulating pathways that contribute to the maintenance of genomic integrity (Fig. 4).

Homologous DNA Recombination

RAD51

Although no physical interaction has been demonstrated between WRN and RAD51, a co-localization study found that WRN forms nuclear foci in response to various DNA damaging agents, and that these WRN foci partially overlap with the foci of RAD51.[13] These results suggest that WRN and RAD51 might function in concert in processes such as HR repair and processing of stalled replication forks (see Figs. 5 and 6).

RAD52

RAD52 is a single strand annealing protein that mediates recombination by stimulating DNA strand exchange reaction and Holliday junction formation. Recently, RAD52 has been shown to physically and functionally interact with WRN both in vitro and in vivo.[46] RAD52 and WRN co-localize to stalled replication forks upon DNA damage with the intra-strand cross-linker mitomycin C.[46] RAD52 inhibits the exonuclease activity of WRN, but stimulates the WRN helicase activity on DNA substrates containing a bubble or a D-loop structure. These results suggest that coordinated WRN and RAD52 activities are involved in replication fork rescue after DNA damage (see Figs. 5 and 6).

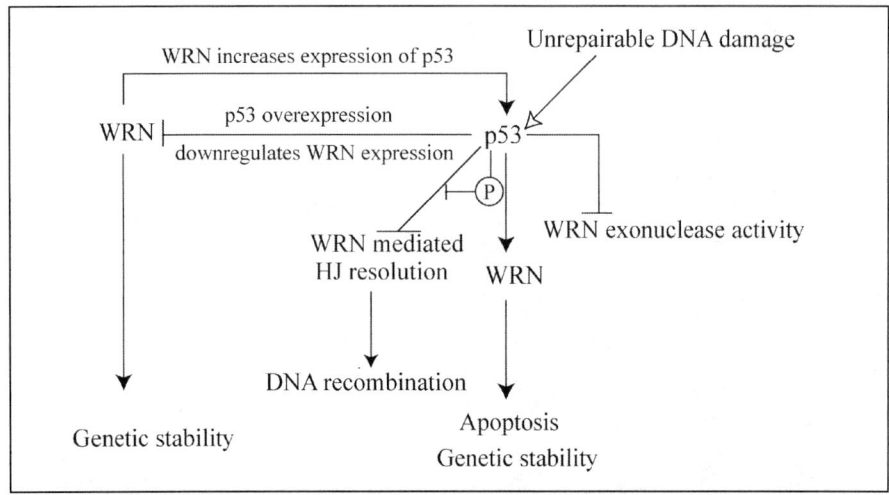

Figure 4. Schematic representation of p53 and WRN interaction and inter-regulation. WRN and p53 inter-regulate their expression levels. WRN functions in the p53 mediated apoptotic pathway but has also p53 independent functions that contribute to genomic stability. In the apoptotic pathway, p53 binds to WRN to inhibit WRN exonuclease activity and WRN ability to disrupt HJ. Phosphorylation of p53 modulates its action on WRN. Sumoylation changes the half-life of p53 and could potentially modulate p53-WRN interaction in the apoptotic pathway, but no information on this topic is available yet.

BLM

BLM is another RecQ helicase which when mutated leads to Bloom syndrome, a highly cancer-prone disease associated with increased genomic instability and elevated sister chromatid exchanges. BLM and WRN co-immunoprecipitate from a nuclear matrix–solubilized fraction, and the recombinant proteins can interact directly. Furthermore, BLM and WRN co-localize to nuclei foci in several different human cell lines.[47] There is no interaction or synergism between WRN and BLM in the helicase reaction, but BLM can inhibit the exonuclease activity of WRN. While several other proteins were shown to inhibit WRN exonuclease activity, examples of which include p53, DNA-PKcs, RAD52, and c-Abl, the Ku heterodimer dramatically stimulates the exonuclease activity of WRN, and it can do so even in the presence of BLM.[47] Thus, WRN exonuclease activity seems to be modulated by interaction with different proteins, but how this interaction is regulated remains a question that has to be yet elucidated.

Non-Homologous End-Joining: DNA-PK and the Ku Complex

In eukaryotic cells, two major pathways are available to repair double-strand breaks: homologous recombination and nonhomologous end joining (NHEJ). DNA-PK is a key component of the mammalian NHEJ repair pathway. The DNA-dependent protein kinase (DNA-PK) consists of three subunits: a catalytic subunit, DNA-PK$_{cs}$, and the Ku complex, which is a heterodimer comprising Ku70 and Ku80 subunits. The Ku complex binds tightly to DNA double-strand breaks in a sequence-independent manner, and when DNA-PK$_{cs}$ associates with Ku bound to DNA termini, the kinase function of DNA-PK is activated.

The Ku Complex

WRN was shown to interact with the Ku complex by affinity chromatography, and the binding was further confirmed by immunoprecipitation.[48,49] The Ku complex was further shown to enhance the 3'-5'exonuclease activity of WRN and to stimulate WRN to degrade various DNA substrates, such as ssDNA, blunt-ended or 3'-protruding double-stand DNA.[49]

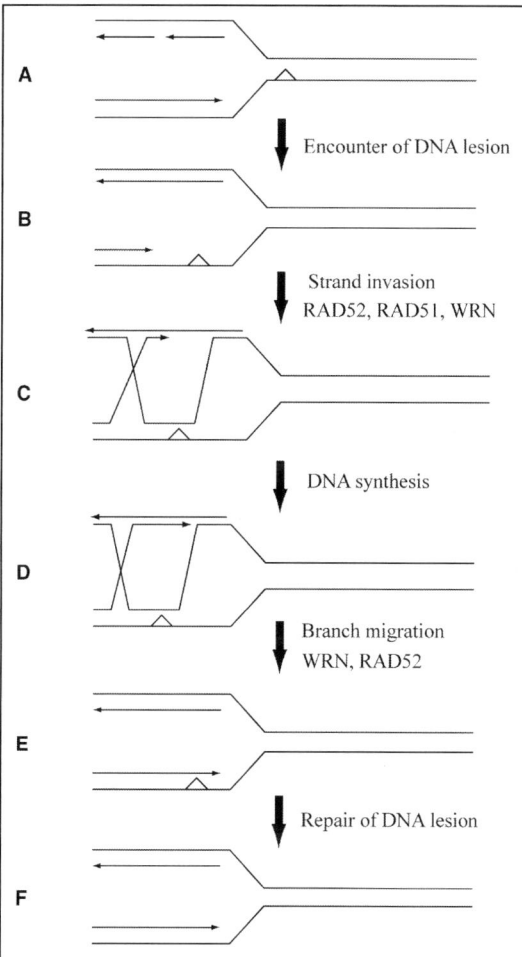

Figure 5. Model for WRN function in initiating strand invasion at a stalled replication fork. A) Replication fork encounters replication block and, B) a gap forms in the stalled fork. C) WRN might create a single-stranded region to promote RAD51-mediated strand-invasion and D-Loop formation. D) DNA synthesis. E) After Holliday junction resolution, the DNA lesion can be removed by an established DNA repair process.

In a study aimed to examine whether specific DNA damaging agents have any blocking effect on the exonuclease activity of WRN, it was found that uracil, hypoxanthine, and ethenoadenine do not present any hindrance for the exonuclease, whereas certain oxidative lesions such as 8-oxoguanine and bulky adducts cause an arrest of the exonuclease digestion.[50] Interestingly, addition of Ku helps the exonuclease activity of WRN to overcome the blockage imposed by 8-oxoA and 8-oxoG adducts, but is much less effective in promoting exonucleolytic digestion past an apurinic or cholesterol moiety.[51] Moreover, it was observed that while there is partial coincidence of the immunofluorescence of WRN and Ku in the absence of DNA-damaging treatments, the co-localization of WRN and Ku is strongly enhanced by 4NQO treatment.[51] WS cells are hypersensitive to 4NQO and treatment with this carcinogen results in increased oxidative damage and several types of bulky lesions in DNA. These results indicate that the interaction between WRN and Ku could be important in the processing of specific types of DNA damage.

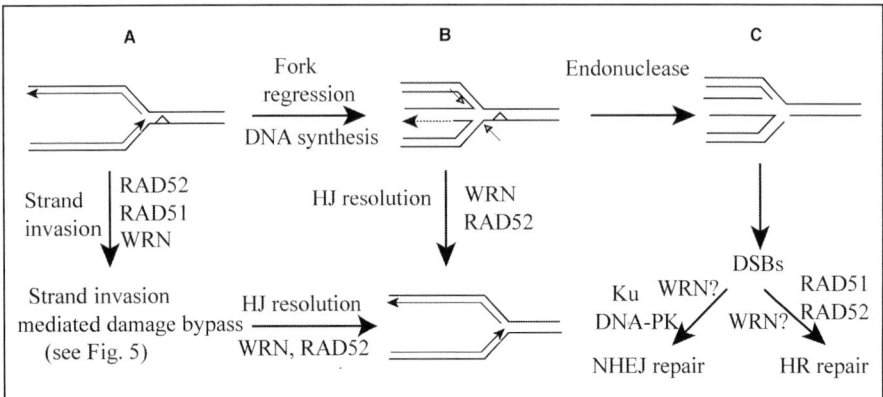

Figure 6. Proposed roles for WRN in HR repair. Stalling of the replication fork can be restored by: A) recombination mediated damage-bypass processes, in which allegedly RAD51 mediates strand invasion, and WRN in association with RAD52 promotes D-loop migration and HJ resolution; B) fork regression creating a chicken-foot intermediate in which leading strand polymerases bypasses the lesion when fork regression is reversed. WRN functions in promoting reversion of regressed fork; C) The chicken foot can be cleaved by HJ resolvases or endonucleases, leading to strand breaks, which can be repaired by HR mechanisms or by error-prone mechanisms such as NHEJ. It is currently unknown whether WRN is involved in DSB repair.

DNA-PK$_{cs}$

The catalytic subunit of the DNA-PK, DNA-PK$_{cs}$, was shown to interact directly with WRN by co-immunoprecipitation and reconstituted interaction using purified proteins and DNA.[52] Ku does not seem to compete or disrupt the DNA-PK$_{cs}$-WRN complex. The assembly of a WRN-DNA-PK-Ku complex was also demonstrated, and this complex appears to be more stable than sub-complexes, suggesting that the WRN-Ku and DNA-PK$_{cs}$-WRN interactions are additive, and not competitive, with respect to WRN. Surprisingly, the consequence of WRN-DNA-PK$_{cs}$ interaction is opposite to the effect of the WRN-Ku interaction. DNA-PK$_{cs}$ was found to dramatically inhibit WRN helicase activity, whereas Ku had little effect on the helicase activity of WRN; Ku stimulated WRN exonuclease activity, whereas DNA-PK$_{cs}$ inhibited the same activity. Most importantly, however, addition of Ku relieved the DNA-PKcs inhibition of both WRN enzymatic functions. Finally, WS cells were found to be mildly sensitive to ionizing radiation,[52] and to cross-linking agents that induce double-strand breaks, suggesting that WRN might play a role in double-strand break repair. However, the activity of WRN on abnormal ends that form during strand-breaking processes has not yet been shown, and the requirement for WRN in double-strand break repair remains to be demonstrated (see also Fig. 6).

DNA-PK$_{cs}$-Mediated Phosphorylation of WRN

DNA-PK was reported to phosphorylate WRN in vitro and to be required for WRN phosphorylation in vivo.[52] A different study found that WRN is phosphorylated in vivo after treatment of cells with DNA damaging agents, and that the level of phosphorylation appears to regulate WRN catalytic activities.[53] More specifically, it was found that treatment of recombinant WRN with a Ser/Thr phosphatase enhances WRN exonuclease and helicase activities, while rephosphorylation of WRN with DNA-PK inhibits its catalytic activities.[53] These results establish a physiological interaction between DNA-PK and WRN, and suggest that WRN phosphorylation by DNA-PK may be a way of regulating its different catalytic activities.

Protein Degradation-p97/VCP

p97/VCP is an AAA ATPase enzyme that mediates several functions in eukaryotic cells, including protein folding and ubiquitin-dependent degradation.[54] VCP was shown to physically interact with WRN in the nucleoli, and the nucleolar VCP/WRN complex was shown to dissociate in response camptothecin, an inhibitor of Topoisomerase I.[55] These results raise the possibility that VCP modulates WRN availability during DNA damage response.

Post-Translational Modification of WRN and Its Localization

In response to DNA damage, DNA-PK phosphorylates the nuclear tyrosine kinase c-Abl,[56] which is another regulator of the DNA damage response. It was found that tyrosine phosphorylation of WRN protein by c-Abl causes the translocation of WRN from nucleoli to discrete nuclear foci upon bleomycin treatments.[57] Interestingly, WRN is not found in nucleoli of chronic myeloid leukemia cells which contain a BCR-Abl fusion protein. BCR-Abl is known to have a constitutive kinase activity. Tyrosine phosphorylation of WRN inhibits both its exonuclease and helicase activities in vitro. These findings suggest another signaling pathway by which c-Abl mediates WRN nuclear localization and catalytic activities in response to DNA damage.[57]

In addition to being phosphorylated, WRN is also acetylated in vivo. This is markedly stimulated by transfection of the acetyltransferase p300 cDNA.[58] Acetylation causes WRN to translocate from nucleoli to specific nuclear areas, and p300 augments this translocation. Furthermore, the protein deacetylase inhibitor, Trichostatin A, enhances WRN translocation to nucleoplasmic foci and delays the re-entry of WRN into the nucleolus at late times after irradiation.[58]

Finally, WRN was shown to be covalently attached with SUMO-1 via the conjugating enzyme Ubc9.[59] The UBC9/SUMO-1 pathway plays a role in cell cycle transition,[60] and is also linked to regulation of intranuclear localization of certain proteins.[61] While no reports are available yet, it is possible that sumoylation might affect the availability or localization of WRN (see next section).

Taken together, these results indicate that several types of post-translational modifications modulate the localization and activities of WRN protein upon DNA damage.

UBC9-Dependent SUMO-1-Modification of WRN

Ubc9 conjugates SUMO-1 to target proteins in a biochemical pathway related to that of ubiquitination. Ubc9 and SUMO-1 were found to interact with WRN by the yeast two-hybrid method, and covalent modification of WRN by SUMO-1 was further confirmed by immunoprecipitation.[59] In addition to WRN, p53, PCNA, and DNA topoisomerase I were shown to be SUMO-1 modified. This interaction seems to be conserved across species, since we have found that also the budding yeast homologue of WRN, Sgs1, interacts with yeast SUMO-1, and its function is highly required in *ubc9-1* mutants in the presence of DNA damage or replication stress (D. Branzei et al, submitted manuscript). We also found that the UBC9/SUMO-1 pathway might play a role in modulating replication by inhibiting ubiquitin-mediated replication fork restart and by sumoylating different proteins at the replication fork (D. Branzei et al, submitted manuscript). Potentially, SUMO-1 modification of WRN, replication proteins, and p53 could be critical in modulating DNA replication and repair, and thus in regulating pathways that contribute to maintenance of genomic stability.

WRN Functions

Resolving Aberrant DNA Structures?

WRN protein can resolve aberrant structures such as G4 DNA,[62] Holliday junctions,[7] synthetic 3- or 4-way junction DNA and extra-helical loop structures,[63] suggesting that WRN unwinds aberrant DNA structures that might impede DNA synthesis or other processes of

DNA metabolism. The fact that cells from WS patients are not sensitive to a variety of DNA damaging agents suggests that WRN is not involved in DNA repair per se, but rather functions to resolve aberrant DNA structures and thus to facilitate DNA transactions. Several lines of evidence come in support of this hypothesis: (A) WRN was found to interact with DNA structure resolution-related proteins such as topoisomerase I,[9] topoisomerase II,[15] the Ku complex,[48] and BLM;[47] (B) WS cells are sensitive to 4NQO[64] and the topoisomerase inhibitor camptothecin,[4] which are believed to cause DNA strand breaks and altered DNA structures in cells; (C) WRN interaction with RPA, RAD52 and p53 affect its ability to process complex DNA structures,[10-12,46] or its ability to disrupt HJ.[41]

While it is clear that WRN can disrupt different types of aberrant DNA structures, it remains to be elucidated whether such structures accumulate in WS cells and whether deficits in resolving them are of pathological significance.

Processing Stalled Replication Forks to Promote Replication Restart?

Model 1. WRN Might Help to Stabilize the Fork Structure by Disrupting 3- or 4-Way Junction DNA

Aberrant DNA structures such as 3- or 4-way junction are known to result in collapse of the replication machinery. It is likely that WRN functions to resolve these structures, and therefore stabilizes the replication forks by preventing aberrant recombination to occur at stalled forks.

Model 2. WRN Might Promote Degradation of Okazaki Fragments on the Lagging Strand at Stalled Replication Forks

Studies in *E. coli* have shown that extensive degradation of the nascent DNA occurs in the absence of the *rec* genes, *recA*, *recF*, or *recR* genes, suggesting that these genes are required to protect and maintain replication forks that are arrested at DNA lesions. In characterizing the nascent DNA degradation that occurs after UV-irradiation, it was demonstrated that other *recF* pathway genes, *recJ* and *recQ*, are responsible for the degradation, and that this degradation seems to occur preferentially on the nascent lagging strand of the blocked replication forks before their resumption. Courcelle and Hanawalt proposed a model in which RecQ and RecJ selectively degrade the nascent lagging strand and thus lengthen the single-strand region at the replication fork.[26] This single-strand structure is then stabilized to a triple-stranded structure consisting of replicated duplex leading strand and unreplicated lagging strand by RecA- mediated homologous strand pairing. This action would allow time for the repair of the downstream DNA lesion and would help prevent recombination from occurring during DNA replication. Potentially reflective of this role, RecQ and Sgs1 have recently been shown to reduce the frequency of illegitimate recombination in *E.coli* and *S. cerevisiae*. Furthermore, studies in *S. cerevisiae* have shown that *WRN* (and *BLM*) can suppress the hyper-recombination (both homologous and illegitimate) phenotype displayed by *sgs1* mutants.[65]

Based on the model proposed by Courcelle and Hanawalt, we speculate that WRN might also promote degradation of Okazaki fragments on the lagging strand of stalled replication forks in association with a functional homologue of RecJ, a 5'-3' ssDNA exonuclease (Fig. 3). The RecJ function in human cells could be achieved by an as yet unidentified 5'-3' ssDNA exonuclease, or by FEN-1, the cleavage activity of which is stimulated by WRN with which it physically interacts.[20] Degradation of the lagging strand would permit stabilization of the replication fork to a triple-stranded structure, in a process mediated perhaps by RAD51, a human RecA homologue. It is therefore interesting that WRN was found to partially co-localize with RAD51 when double-strand breaks and stalled replication forks are induced.[13] Thus, it is possible that WRN forms a complex with Rad51 or functions cooperatively with Rad51 to stabilize the triple-stranded structure in the fork prior to removal of DNA damage (Fig. 3).

Model 3. WRN Might Function in Strand Invasion Mediated Repair after Replication Fork Stalling

In *E. coli*, in a *recBCsbcBC* background in which the major RecBCD recombinational pathway is inactive, the RecF pathway (including the *recF*, *recO*, *recR*, *recQ*, and *recJ* genes) is required for both homologous recombination and recombinational repair. Although the role of RecQ in these processes remains to be established, the observation that RecQ can initiate and disrupt homologous DNA pairing in concert with RecA and SSB[66] indicates that RecQ helicases might be involved. It is therefore conceivable that WRN helicase/exonuclease might create a single-stranded DNA segment in the stalled fork to facilitate strand invasion, and at a second stage might associate with RAD52, RAD51 and RPA to promote D-loop formation, branch migration and/or joint molecular resolution (Fig. 5).

Other Functions of WRN in DNA Replication?

Beside its roles in restoring stalled replication forks, several lines of evidence suggest that WRN might be involved in other processes of DNA replication as well. For instance, FFA-1, the homologue of WRN in *Xenopus laevis*, was shown to be required for formation of replication foci on sperm chromatin.[21] Although no such role has yet been demonstrated for WRN, an early study reports that WS cells show a reduced frequency of replicon initiation,[5] indicating that WRN might be actually involved in initiation of DNA replication.

The functional and physical interactions between WRN and DNA polymerase δ, as well as the physical interactions with PCNA, WRNIP1, and topoisomerase I, suggest that WRN is not merely involved in clearing the path for the replication machinery by resolving aberrant structures that would otherwise hinder the progression of the replication fork associated with the replication machinery, but is implicated in replication, perhaps by stabilizing the replication complex and/or by promoting replication restart pathways at sites where the DNA replication fork has stalled.

Involved in NHEJ?

The physical and functional interactions between WRN and DNA-PK$_{cs}$ and the Ku complex suggest that WRN might be involved in NHEJ in concert with these proteins (see Fig. 6). Although these data suggest that WRN might function in a subset of NHEJ events, and connections between HR and NHEJ have been demonstrated,[67] the activity of WRN on abnormal ends that form during strand-breaking processes has not yet been demonstrated, and the requirement for WRN in double-strand break repair remains an issue open to debate. Furthermore, the resemblance between WS phenotypes and those of other mammalian HR mutants[68] suggests that even if WRN plays a role in NHEJ, that is probably subservient to its role in HR.

Involved in Homologous Recombination Repair?

A study aimed to determine the role that *WRN* might play in recombination in human somatic cells, found that WS and control cell lines initiate mitotic recombination at comparable rates, but WS cells fail to resolve recombinant products to yield viable progeny.[69] Furthermore, molecular analysis of mitotic recombinants in WS cells that retained growth potential found a decrease in conversion-type recombinants in these cells, indicating a role for WRN in mitotic recombination.[69] A role for WRN in promoting HR in vertebrate cells comes also from analysis of sister chromatid exchange (SCE) frequency in $BLM^{-/-}WRN^{-/-}$ DT40 cells. Similar to Bloom syndrome (BS) cells, $BLM^{-/-}$ DT40 cells show increased frequencies in SCEs, which have been demonstrated to occur via HR in a Rad54 dependent manner.[70,71] Interestingly, disruption of WRN partially diminished the increased SCE frequency of $BLM^{-/-}$ DT40 cells,[72] suggesting that WRN contributes to and accelerates HR repair in vertebrate cells.

Finally, a role for WRN in the resolution of HR products, in general, and of HJ, in particular, was revealed by demonstrating that (A) a dominant-negative form of mammalian RAD51 protein that suppresses HR in WS and control cells improved survival of WS cells after DNA damage; and (B) that expression active RusA protein, a bacterial resolvase that binds a variety of DNA junction structures but can efficiently cleave only four-way Holliday junctions, can significantly improve WS cell survival and recombination phenotypes (see Fig. 6).[73]

These results indicate that WRN is likely to participate in resolving recombination intermediates that result from RAD51 action, a portion of which contain HJ (Fig. 6). Failure to do so is likely to lead to mitotic arrest and cell death when recognized by programmed cell death and apoptosis pathways, or lead to genomic instability when captured by error-prone pathways such as NHEJ.

Concluding Remarks

WRN is one of the five already characterized RECQ helicases in human cells, and mutations in three of these RECQ helicases were found to lead to genetic disorders characterized by genomic instability. In addition, Werner syndrome patients show premature aging. While the exact role of WRN in DNA metabolism is not yet clear, biochemical studies suggest that WRN probably shares some of its responsibilities with other RECQ helicases, such as BLM. However, the functions of these two helicases are clearly not redundant, since mutations in WRN or BLM alone lead to independent genetic disorders. Therefore, a challenging topic that remains to be tackled with is to uncover the unique functions of WRN among RecQ helicases, and find out how mutations in a single gene can lead to such a wide array of phenotypes, and in particular to the early-onset of age-related pathologies.

The biochemical characterization of WRN showed that WRN helicase/exonuclease deals with alternate DNA structures that form during DNA replication or as a result of DNA damage. However, it is still not known whether these activities function simultaneously or separately, and whether interacting proteins are required to regulate the timing of the two activities, and the processes into which WRN participates. While these questions remain to be answered, it is clear that the proteins found to interact with WRN are involved in processes of DNA metabolism similar to the ones suspected for WRN, such as DNA replication and recombination, and that many of them are able to regulate WRN exonuclease and helicase activities, WRN availability and localization, or WRN functions in processes of DNA metabolism. We believe that by understanding the molecular basis of these interactions, we will be able to better understand the function of WRN at the replication fork and in maintaining genomic stability. Without doubt, the knowledge derived from these studies will not only illuminate the molecular mechanism of WS, but will also serve to understand the molecular mechanisms at the replication fork in the ensuing years.

References

1. Goto M, Yamabe Y, Shiratori M et al. Immunological diagnosis of Werner syndrome by down-regulated and truncated gene products. Hum Genet 1999; 105:301-307.
2. Fujiwara Y, Higashikawa T, Tatsumi M. A retarded rate of DNA replication and normal level of DNA repair in Werner's syndrome fibroblasts in culture. J Cell Physiol 1977; 92:365-374.
3. Takeuchi F, Hanaoka F, Goto M et al. Prolongation of S phase and whole cell cycle in Werner's syndrome fibroblasts. Exp Gerontol 1982; 17:473-480.
4. Poot M, Hoehn H, Runger TM et al. Impaired S-phase transit of Werner syndrome cells expressed in lymphoblastoid cell lines. Exp Cell Res 1992; 202:267-273.
5. Takeuchi F, Hanaoka F, Goto M et al. Altered frequency of initiation sites of DNA replication in Werner's syndrome cells. Hum Genet 1982; 60:365-368.
6. Yan H, Chen CY, Kobayashi R et al. Replication focus-forming activity 1 and the Werner syndrome gene product. Nat Genet 1998; 19:375-378.
7. Constantinou A, Tarsounas M, Karow JK et al. Werner's syndrome protein (WRN) migrates Holliday junctions and co- localizes with RPA upon replication arrest. EMBO Rep 2000; 1:80-84.

8. Lebel M, Leder P. A deletion within the murine Werner syndrome helicase induces sensitivity to inhibitors of topoisomerase and loss of cellular proliferative capacity. Proc Natl Acad Sci USA 1998; 95:13097-13102.
9. Lebel M, Spillare EA, Harris CC et al. The Werner syndrome gene product co-purifies with the DNA replication complex and interacts with PCNA and topoisomerase I. J Biol Chem 1999; 274:37795-37799.
10. Shen JC, Gray MD, Oshima J et al. Characterization of Werner syndrome protein DNA helicase activity: directionality, substrate dependence and stimulation by replication protein A. Nucleic Acids Res 1998; 26:2879-2885.
11. Brosh RM Jr, Orren DK, Nehlin JO et al. Functional and physical interaction between WRN helicase and human replication protein A. J Biol Chem 1999; 274:18341-18350.
12. Ohsugi I, Tokutake Y, Suzuki N et al. Telomere repeat DNA forms a large non-covalent complex with unique cohesive properties which is dissociated by Werner syndrome DNA helicase in the presence of replication protein A. Nucleic Acids Res 2000; 28:3642-3648.
13. Sakamoto S, Nishikawa K, Heo SJ et al. Werner helicase relocates into nuclear foci in response to DNA damaging agents and co-localizes with RPA and Rad51. Genes Cells 2001; 6:421-430.
14. Jonsson ZO, Hubscher U. Proliferating cell nuclear antigen: more than a clamp for DNA polymerases. Bioessays 1997; 19:967-975.
15. Franchitto A, Oshima J, Pichierri P. The G2-phase decatenation checkpoint is defective in Werner syndrome cells. Cancer Res 2003; 63:3289-3295.
16. Szekely AM, Chen YH, Zhang C et al. Werner protein recruits DNA polymerase delta to the nucleolus. Proc Natl Acad Sci USA 2000; 97:11365-11370.
17. Kamath-Loeb AS, Johansson E, Burgers PM et al. Functional interaction between the Werner Syndrome protein and DNA polymerase delta. Proc Natl Acad Sci USA 2000; 97:4603-4608.
18. Kamath-Loeb AS, Loeb LA, Johansson E et al. Interactions between the Werner syndrome helicase and DNA polymerase delta specifically facilitate copying of tetraplex and hairpin structures of the d(CGG)n trinucleotide repeat sequence. J Biol Chem 2001; 276:16439-16446.
19. Harrigan JA, Opresko PL, von Kobbe C et al. The Werner syndrome protein stimulates DNA polymerase beta strand displacement synthesis via its helicase activity. J Biol Chem 2003; 278:22686-22695.
20. Brosh RM Jr, von Kobbe C, Sommers JA et al. Werner syndrome protein interacts with human flap endonuclease 1 and stimulates its cleavage activity. Embo J 2001; 20:5791-5801.
21. Yan H, Newport J. FFA-1, a protein that promotes the formation of replication centers within nuclei. Science 1995; 269:1883-1885.
22. Chen CY, Graham J, Yan H. Evidence for a replication function of FFA-1, the Xenopus orthologue of Werner syndrome protein. J Cell Biol 2001; 152:985-996.
23. Wilson DM 3rd, Carney JP, Coleman MA et al. Hex1: a new human Rad2 nuclease family member with homology to yeast exonuclease 1. Nucleic Acids Res 1998; 26:3762-3768.
24. Tishkoff DX, Amin NS, Viars CS et al. Identification of a human gene encoding a homologue of Saccharomyces cerevisiae EXO-1, an exonuclease implicated in mismatch repair and recombination. Cancer Res 1998; 58:5027-5031.
25. Sharma S, Sommers JA, Driscoll HC et al. The exonucleolytic and endonucleolytic cleavage activities of human exonuclease 1 are stimulated by an interaction with the carboxyl-terminal region of the Werner syndrome protein. J Biol Chem 2003; 278:23487-23496.
26. Courcelle J, Hanawalt PC. RecQ and RecJ process blocked replication forks prior to the resumption of replication in UV-irradiated Escherichia coli. Mol Gen Genet 1999; 262:543-551.
27. Kawabe Y, Branzei D, Hayashi T et al. A novel protein interacts with the Werner's syndrome gene product physically and functionally. J Biol Chem 2001; 276:20364-20369.
28. Hishida T, Iwasaki H, Ohno T et al. A yeast gene, MGS1, encoding a DNA-dependent AAA(+) ATPase is required to maintain genome stability. Proc Natl Acad Sci USA 2001; 98:8283-8289.
29. Branzei D, Seki M, Onoda F et al. The product of Saccharomyces cerevisiae WHIP/MGS1, a gene related to replication factor C genes, interacts functionally with DNA polymerase delta. Mol Genet Genomics 2002; 268:371-86.
30. Barre FX, Soballe B, Michel B et al. Circles: the replication-recombination-chromosome segregation connection. Proc Natl Acad Sci USA 2001; 98:8189-8195.
31. Branzei D, Seki M, Onoda F et al. Characterization of the slow-growth phenotype of S. cerevisiae Whip/Mgs1 Sgs1 double deletion mutants. DNA Repair (Amst) 2002; 1:671-682.
32. Hishida T, Ohno T, Iwasaki H et al. Saccharomyces cerevisiae MGS1 is essential in strains deficient in the RAD6-dependent DNA damage tolerance pathway. Embo J 2002; 21:1-11.
33. Trucco C, Oliver FJ, de Murcia G et al. DNA repair defect in poly(ADP-ribose) polymerase-deficient cell lines. Nucleic Acids Res 1998; 26:2644-9.

34. Le Rhun Y, Kirkland JB, Shah GM. Cellular responses to DNA damage in the absence of Poly(ADP-ribose) polymerase. Biochem Biophys Res Commun 1998; 245:1-10.
35. Lebel M, Lavoie J, Gaudreault I et al. Genetic cooperation between the Werner syndrome protein and poly(ADP-ribose) polymerase-1 in preventing chromatid breaks, complex chromosomal rearrangements, and cancer in mice. Am J Pathol 2003; 162:1559-69.
36. Menissier de Murcia J, Ricoul M, Tartier L et al. Functional interaction between PARP-1 and PARP-2 in chromosome stability and embryonic development in mouse. Embo J 2003; 22:2255-63.
37. Blander G, Kipnis J, Leal JF et al. Physical and functional interaction between p53 and the Werner's syndrome protein. J Biol Chem 1999; 274:29463-29469.
38. Spillare EA, Robles AI, Wang XW et al. p53-mediated apoptosis is attenuated in Werner syndrome cells. Genes Dev 1999; 13:1355-1360.
39. Wang XW, Vermeulen W, Coursen JD et al. The XPB and XPD DNA helicases are components of the p53-mediated apoptosis pathway. Genes Dev 1996; 10:1219-1232.
40. Brosh RM Jr, Karmakar P, Sommers JA et al. p53 Modulates the exonuclease activity of Werner syndrome protein. J Biol Chem 2001; 276:35093-102.
41. Yang Q, Zhang R, Wang XW et al. The processing of Holliday junctions by BLM and WRN helicases is regulated by p53. J Biol Chem 2002; 277:31980-7.
42. Lebel M, Cardiff RD, Leder P. Tumorigenic effect of nonfunctional p53 or p21 in mice mutant in the Werner syndrome helicase. Cancer Res 2001; 61:1816-1819.
43. Oshima J, Campisi J, Tannock TC et al. Regulation of c-fos expression in senescing Werner syndrome fibroblasts differs from that observed in senescing fibroblasts from normal donors. J Cell Physiol 1995; 162:277-283.
44. Blander G, Zalle N, Leal JF et al. The Werner syndrome protein contributes to induction of p53 by DNA damage. Faseb J 2000; 14:2138-2140.
45. Yamabe Y, Shimamoto A, Goto M et al. Sp1-mediated transcription of the Werner helicase gene is modulated by Rb and p53. Mol Cell Biol 1998; 18:6191-6200.
46. Baynton K, Otterlei M, Bjoras M et al. WRN interacts physically and functionally with the recombination mediator protein RAD52. J Biol Chem 2003; 278:36476-36486.
47. von Kobbe C, Karmakar P, Dawut L et al. Colocalization, physical, and functional interaction between Werner and Bloom syndrome proteins. J Biol Chem 2002; 277:22035-22044.
48. Cooper MP, Machwe A, Orren DK et al. Ku complex interacts with and stimulates the Werner protein. Genes Dev 2000; 14:907-912.
49. Li B, Comai L. Functional interaction between Ku and the Werner syndrome protein in DNA end processing. J Biol Chem 2000; 275:39800.
50. Bohr VA, Cooper M, Orren D et al. Werner syndrome protein: biochemical properties and functional interactions. Exp Gerontol 2000; 35:695-702.
51. Orren DK, Machwe A, Karmakar P et al. A functional interaction of Ku with Werner exonuclease facilitates digestion of damaged DNA. Nucleic Acids Res 2001; 29:1926-1934.
52. Yannone SM, Roy S, Chan DW et al. Werner syndrome protein is regulated and phosphorylated by DNA- dependent protein kinase. J Biol Chem 2001; 276:38242-38248.
53. Karmakar P, Piotrowski J, Brosh RM Jr et al. Werner protein is a target of DNA-PK in vivo and in vitro and its catalytic activities are regulated by phosphorylation. J Biol Chem 2002; 277:18291-18302.
54. Wang Q, Song C, Li CC. Hexamerization of p97-VCP is promoted by ATP binding to the D1 domain and required for ATPase and biological activities. Biochem Biophys Res Commun 2003; 300:253-260.
55. Partridge JJ, Lopreiato JO Jr, Latterich M et al. DNA Damage Modulates Nucleolar Interaction of the Werner Protein with the AAA ATPase p97/VCP. Mol Biol Cell 2003; 14:4221-4229.
56. Kharbanda S, Pandey P, Jin S et al. Functional interaction between DNA-PK and c-Abl in response to DNA damage. Nature 1997; 386:732-735.
57. Cheng WH, von Kobbe C, Opresko PL et al. Werner syndrome protein phosphorylation by abl tyrosine kinase regulates its activity and distribution. Mol Cell Biol 2003; 23:6385-6395.
58. Blander G, Zalle N, Daniely Y et al. DNA damage-induced translocation of the Werner helicase is regulated by acetylation. J Biol Chem 2002; 277:50934-50940.
59. Kawabe Y, Seki M, Seki T et al. Covalent modification of the Werner's syndrome gene product with the ubiquitin-related protein, SUMO-1. J Biol Chem 2000; 275:20963-20966.
60. Seufert W, Futcher B, Jentsch S. Role of a ubiquitin-conjugating enzyme in degradation of S- and M-phase cyclins. Nature 1995; 373:78-81.
61. Ishov AM, Sotnikov AG, Negorev D et al. PML is critical for ND10 formation and recruits the PML-interacting protein Daxx to this nuclear structure when modified by SUMO-1. J Cell Biol 1999; 147:221-234.

62. Fry M, Loeb LA. Human werner syndrome DNA helicase unwinds tetrahelical structures of the fragile X syndrome repeat sequence d(CGG)n. J Biol Chem 1999; 274:12797-12802.
63. Shen J, Loeb LA. Unwinding the molecular basis of the Werner syndrome. Mech Ageing Dev 2001; 122:921-944.
64. Ogburn CE, Oshima J, Poot M et al. An apoptosis-inducing genotoxin differentiates heterozygotic carriers for Werner helicase mutations from wild-type and homozygous mutants. Hum Genet 1997; 101:121-125.
65. Yamagata K, Kato J, Shimamoto A et al. Bloom's and Werner's syndrome genes suppress hyperrecombination in yeast sgs1 mutant: implication for genomic instability in human diseases. Proc Natl Acad Sci USA 1998; 95:8733-8738.
66. Harmon FG, Kowalczykowski SC. RecQ helicase, in concert with RecA and SSB proteins, initiates and disrupts DNA recombination. Genes Dev 1998; 12:1134-1144.
67. Richardson C, Jasin M. Coupled homologous and nonhomologous repair of a double-strand break preserves genomic integrity in mammalian cells. Mol Cell Biol 2000; 20:9068-9075.
68. van Gent DC, Hoeijmakers JH, Kanaar R. Chromosomal stability and the DNA double-stranded break connection. Nat Rev Genet 2001; 2:196-206.
69. Prince PR, Emond MJ, Monnat RJ Jr. Loss of Werner syndrome protein function promotes aberrant mitotic recombination. Genes Dev 2001; 15:933-938.
70. Sonoda E, Sasaki MS, Morrison C et al. Sister chromatid exchanges are mediated by homologous recombination in vertebrate cells. Mol Cell Biol 1999; 19:5166-5169.
71. Wang W, Seki M, Narita Y et al. Possible association of BLM in decreasing DNA double strand breaks during DNA replication. Embo J 2000; 19:3428-3435.
72. Imamura O, Fujita K, Itoh C et al. Werner and Bloom helicases are involved in DNA repair in a complementary fashion. Oncogene 2002; 21:954-963.
73. Saintigny Y, Makienko K, Swanson C et al. Homologous recombination resolution defect in Werner syndrome. Mol Cell Biol 2002; 22:6971-6978.

CHAPTER 5

Sensitivity of Werner's Syndrome Cells to DNA Damaging Agents: Insights into the Biological Functions of the Werner Protein

Adayabalam S. Balajee and Fabrizio Palitti

Abstract

Werner's syndrome (WS) is a human autosomal recessive disorder characterized by many symptoms of accelerated aging. The gene responsible for Werner's syndrome (*WRN*) has been cloned and the protein has been biochemically characterized as a helicase/exonuclease. Determination of the preferred DNA substrates for WRN helicase and exonuclease has given great insights into the possible biological functions of the *WRN* gene. The extreme sensitivity of WS cells to topoisomerase inhibitors and inter-strand cross-linking agents indicate that the WRN helicase/exonuclease may participate in specific pathways of DNA repair. Interaction of the WRN enzyme with proteins involved in DNA replication, transcription, repair and recombination attests a role for this gene in diverse DNA metabolic activities. This review attempts to unravel the biological function(s) of the WRN helicase/exonuclease on the basis of sensitivity of WS cells to different DNA damaging agents.

Introduction

Cancer incidence is found at a relatively high frequency in certain human population groups for various reasons. At least in some cases, cancer incidence is linked to known autosomal recessive human disorders due to mutations in genes involved in the maintenance of genomic stability. Some of these disorders include xeroderma pigmentosum (XP), Cockayne's syndrome (CS), Werner's syndrome (WS), Bloom's syndrome (BS) and ataxia telangiectasia (AT). Cells from these patients show differential sensitivity to many chemical and physical agents due to defects in specific DNA repair pathways. The correlation between DNA repair defects and the aetiology of these disorders may not be strong in all cases as repair deficiency may vary among the patients due to the mutational status of the responsible genes. Furthermore, identification of the lesion that is chiefly responsible for the cytotoxicity is rather difficult as many DNA damaging agents produce a spectrum of lesions, which require the coordinated activities of multiple repair pathways for their removal. Despite these limitations, evaluation of the cytotoxicity of cells to different DNA damaging agents has often enabled in deducing the biological functions of the gene(s) in question. For example, observation of extreme sensitivity of XP patients to sunlight exposure led to the elucidation of various XP genes and their concerted role in nucleotide excision repair pathway.

The types of DNA lesions induced by various agents have been previously reviewed.[1] Ionizing radiation and radio-mimetic agents such as bleomycin induce single strand breaks, double strand breaks, alkali labile sites, base damage and DNA-protein cross-links. Ultraviolet (UV) light induces pyrimidine dimers of the cyclobutane type, 6-4 photoproducts and DNA-protein cross-links. Alkylated bases, phosphotriesters, inter- and intra-strand cross-links are generated by mono- and poly-functional alkylating agents. In addition to the above agents that cause direct damage to DNA, many chemicals indirectly affect DNA integrity through their interference with DNA replication, transcription, repair and recombination. Some of them include inhibitors of DNA replication (aphidicolin, cytosine arabinoside and hydroxyurea), transcription (α-amanitin), repair (aphidicolin and caffeine) and chromosome condensation (topoisomerase inhibitors such as VP16 and camptothecin). DNA damaging agents can also be categorised depending on their potential to induce chromosome or chromatid type of aberrations in an S phase dependent and independent manner. Ionizing radiation (IR), which induces single and double strand breaks, is a classic example of S phase independent agent. Irradiation of cells in G_1 and G_2 phases results in the formation of chromosome type and chromatid type of aberrations respectively. Unlike IR, S phase is required for the transformation of UV light-induced DNA lesions into chromosomal aberrations. It has been proposed that the chromosomal aberrations induced by S phase independent agents are due to misrepair events and those caused by S phase dependent agents are due to error prone replication events.

Cells derived from some of the cancer prone human disorders are characterized by extreme sensitivity to many clastogenic and mutagenic agents. Increased sensitivity to mutagens has been demonstrated for cells derived from xeroderma pigmentosum (XP), Fanconi anaemia (FA), ataxia telangiectasia (AT) and Bloom's syndrome (BS). FA, AT and BS patients are also characterized by an increase in the spontaneous frequency of chromosomal aberrations. While AT cells are characterized by increased chromosome end fusions, BS cells show high frequency of sister chromatid exchanges (SCE). The gene responsible for BS belongs to the *Escherichia coli* RecQ family of helicases and the *BLM* gene product unwinds duplex DNA in the 3'-5' direction. This processivity is enhanced by replication protein A. In addition to FA, AT and BS, cells derived from a human progeroid syndrome, Werner, is also characterized by chromosomal abnormalities and the *WRN* gene product is another member of the RecQ family of helicases. All these disorders are collectively known as "chromosome breakage syndromes" and active research is directed in recent times towards understanding the biological roles of these genes in chromosome integrity.

Cellular Characteristics of Werner's Syndrome

Many signs of accelerated aging such as bilateral cataracts, diabetes mellitus, osteoporosis, arteriosclerosis and trophic ulcers of the legs characterize Werner's syndrome (WS) patients. In addition, WS patients show enhanced risk of developing various neoplasms including different types of sarcomas and carcinomas. Consistent with a premature aging phenotype, WS fibroblasts also exhibit retarded proliferation potential in culture. At the cellular level, WS is characterized by variegated chromosome translocation mosaicism involving the expansion of different structural chromosome rearrangements in different independent clones of the same cell line. A more recent study[2] has confirmed and expanded this finding using fluorescence in situ hybridisation (FISH): an increase in both unstable and stable aberrations was found in lymphoblastoid and fibroblast cells from WS patients. Furthermore, the WS lymphoblastoid cell line K0375 was characterized by the presence of a balanced translocation involving chromosome 5 and 12 in all the cells and 60% of the cells had an additional translocated chromosome 12. Large genomic DNA deletions have also been observed in WS cells. Fukuchi et al[3] found higher frequency of spontaneous mutations in the *HPRT* gene in 6-thioguanine-resistant WS cells and approximately half of them were characterized by deletions greater than 20 kb of DNA at the *HPRT* locus.

Sensitivity of WS Cells to DNA Damaging Agents

Although WS belongs to the "chromosome breakage syndromes", WS cells are not extremely sensitive to many commonly used DNA damaging agents like IR, UV, mono- and polyfunctional alkylating agents. Gebhart et al[4] reported that WS cells are sensitive to the induction of chromosomal aberrations by 4-nitroquinoline-1-oxide (4NQO), a chemical which produces bulky DNA adducts with a mechanism of action similar to UV light. In addition to bulky DNA adducts, 4NQO induces base damage such as 8-oxoguanine. Ogburn et al[5] found that lymphocytes derived from homozygous and heterozygous carriers of Werner mutation are more prone to apoptosis after 4NQO treatment. The sensitivity to 4NQO was observed in both primary[6] and SV40-transformed fibroblasts.[7] The sensitivity of XP cells to 4NQO suggests that 4NQO induced lesions are repaired by the nucleotide excision repair (NER) pathway. Although WS cells are sensitive to 4NQO, purified WRN protein does not show any binding affinity for 4NQO adducts.[8] It is likely that the WRN protein enzyme may play an accessory role in the repair of 4NQO-induced lesions.

Chromosomal radiosensitivity of two WS lymphoblastoid cell lines was studied by Grigorova et al[2] using both Giemsa staining and fluorescence in situ hybridisation (FISH) with DNA probes specific for chromosomes 1,4,5 and 12. Both WS lymphoblastoid cell lines showed only a slight increase in the frequency of aberrant cells and chromosome fragments upon irradiation as compared to cells derived from normal healthy individuals. Furthermore, WS cells does not show, as oppose to some other "cancer prone" syndromes, a higher sensitivity to the induction of chromosome damage by X-rays in the G_2 phase of the cell cycle, suggestive of no defects in the DNA damage-induced G_2 checkpoint response. WS cells behaved normally to hydroxyurea post-treatment compared to normal cells. Such treatment enhances the frequency of X-ray induced chromosomal aberrations.[9] Caffeine is also a drug known to sensitise cells to IR by abrogating the G_2 checkpoint response. Interestingly, although caffeine post-treatment did not abrogate the G_2 arrest imposed by IR or mitomycin C in WS cells, no enhancement in the chromosomal aberrations induced by these agents was observed in WS cells compared to normal cells. The lack of G_2 arrest abrogation by caffeine indicates the requirement of a functional *WRN* gene product in the signal transduction pathway by which caffeine can override the DNA damage induced G_2 checkpoint.

The sensitivity of WS cells to topoisomerase I (camptothecin) and II (VP-16 and amasacrine) inhibitors in S and G_2 phases of the cell cycle has been recently demonstrated.[10-12] Likewise, embryonic stem cells derived from *Wrn* homozygous mutant mice harbouring a disrupted helicase domain are sensitive to topoisomerase inhibitors.[13] Although the basis for the enhanced sensitivity of WS cells to topoisomerase inhibitors remains unclear, the general speculation is that the WRN helicase together with topoisomerases may mediate an effective recombinational repair pathway, operating in the S and G_2 phases of the cell cycle prior to mitosis.

Studies on the budding yeast *Saccharomyces cerevisiae* have also revealed a molecular link between the Sgs1 protein (the orthologue of WRN and BLM helicases) and topoisomerases. Both topoisomerase III mutants and *Sgs1* mutants are characterized by genomic instability due to hyper-recombinations between highly repetitive DNA sequences. This phenotype is suppressed by the wild type *Sgs1* allele in both these mutant strains. Furthermore, the slow growth phenotype observed in budding yeast topoisomerase III mutants can be rescued by Sgs1.[14]

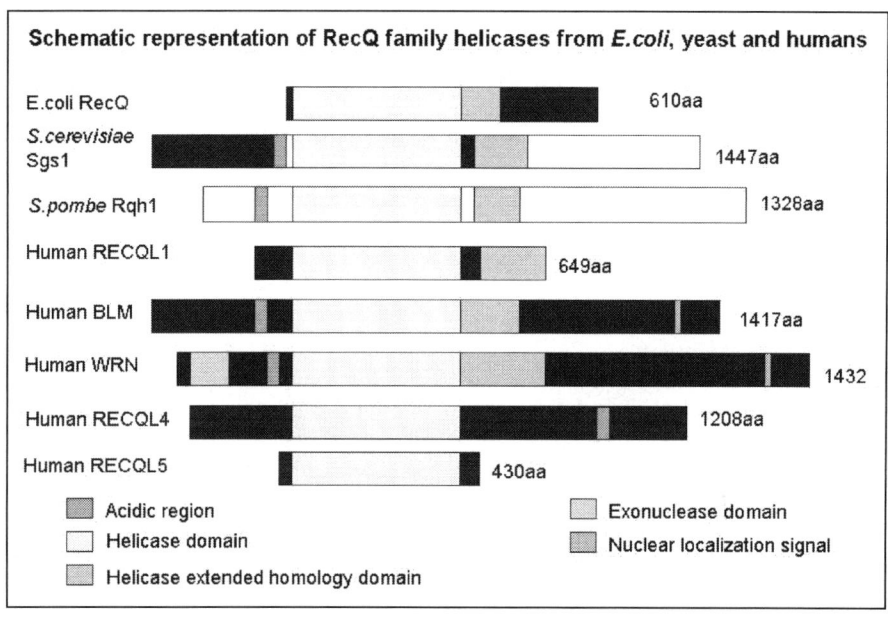

Figure 1. Comparison of regions of homology between different members of the RecQ family of helicases.

Sgs1 has been shown to physically interact with topoisomerases. As Sgs1 physically interacts with topoisomerase III, a likely possibility is that the Sgs1/topoisomerase III complex may resolve topologically altered secondary DNA structures. Watt et al[15] have also demonstrated the importance of an interaction between Sgs1 and topoisomerase II in the maintenance of faithful chromosome segregation. Disruption of the *Sgs1* gene in *S. cerevisiae* results in hypersensitivity to methyl methane sulfonate (MMS), hydroxyurea and leads to a hyper-recombination phenotype involving interchromosomal recombination between heteroallelles.[16] The highly conserved C-terminal region and the conserved helicase motifs of Sgs1 (see Fig. 1) are necessary for the complementation of MMS sensitivity and inter-chromosomal recombinations. Additionally, 45 amino acid residues at the N-terminal region of Sgs1 are required for this complementation. Introduction of mis-sense mutations in the region encoded by amino acid residues four to thirteen completely abolished the ability of *Sgs1* gene to complement the sensitivity of MMS and the hyper-recombination phenotype. The mutational inactivation of this region also prevented the interaction between Sgs1 and topoisomerase III. Thus, these results indicated that the interaction of the N-terminal region of Sgs1 with topoisomerase III is involved in the complementation of the MMS sensitivity and the hyper-recombination features observed in *Sgs1* mutants.[16] This clearly illustrates the importance of the functional interaction between Sgs1 and topoisomerases in genomic stability. Finally, Harmon et al[17,18] showed that the bacterial *E.coli* RecQ helicase can unwind covalently closed double stranded DNA in the presence of single stranded DNA binding (SSB) protein. More importantly, this RecQ unwinding activity also stimulates the *E.coli* topoisomerase III enzyme resulting in the generation of catenated double stranded DNA molecules. This, again, suggests a functional interaction between RecQ helicases and topoisomerase III and their concerted function may regulate recombination processes. A similar functional interaction between the *WRN* gene product and topoisomerases may also exist in mammalian cells.

Increased apoptosis in S phase was observed in WS lymphoblastoid cells upon treatment with melphalan, chlorambucil, mitomycin C and cis-platinum (II) diamine dichloride but not with etoposide, berenil, daunomycin, adriamycin, mitoxantrone, and echinomycin.[19] Of all the drugs tested, mitomycin C, which does not induce any DNA protein cross-links, resulted in a much higher fraction of apoptotic cells than that reported earlier for 4NQO and camptothecin. Camptothecin which induces DNA double strand breaks probably by inhibiting the activity of topoisomerase I activity, elicits an apoptotic response in the S phase of WS cells. In contrast, the topoisomerase II inhibitor etoposide does not result in S phase apoptosis. The difference observed in terms of S phase apoptotic response between topoisomerase I and II inhibitors is difficult to explain as both of these agents induce replication mediated double strand breaks.

Apoptosis was detectable in WS cells early after camptothecin treatment (two to six hours) while apoptosis was observed only after ten hours of camptothecin treatment in normal cells.[20] Identification of S phase cells by bromo-deoxyuridine pulse labeling revealed that some of the apoptotic cells detected during the early times of recovery after camptothecin treatment in WS cells were not in S phase. In fact, the reduction in the number of G2 cells analyzed after treatment indicates that camptothecin-induced apoptosis also occurs in the G2 phase of the cell cycle in WS cells.[20] Interestingly, agents that are expected to block WRN helicase activity such as daunomycin, adriamycin, mitoxantrone, and echinomycin did not elicit the apoptotic response in the S phase of WS cells. Also, agents like berenil that block the access of enzymes into minor groove of the DNA did not cause apoptosis in WS cells.[19] Thus, the selective apoptotic potential of WS cells to DNA damaging agents such as mitomycin C, camptothecin and 4NQO points to a requirement for the coordinated activities of WRN helicase/exonuclease in either the repair or the bypass of certain specific replication blocking lesions like inter-strand cross-links, double strand breaks and bulky DNA adducts.

It is worth mentioning that topoisomerase II inhibitors (VP-16 and amasacrine) elevated the yield of chromosomal aberrations in WS cells specifically in the G2 phase of the cell cycle.[10] Corroborating with this, potentiation of X-ray-induced chromosomal damage by catalytic inhibition of topoisomerase II was also noticed in WS cells only in G2 phase.[12] In contrast to topoisomerase II inhibitors, the topoisomerase I inhibitor camptothecin induces a much higher level of chromosomal damage in both the S and G2 phases of WS cells compared to normal cells.[11] The increased chromosomal damage by topoisomerase II inhibitors in G2 phase of WS cells may be due to error prone replication coupled with the failure to undergo apoptosis in S phase. In any case, the chromosomal sensitivity of WS cells to topoisomerase I and topoisomerase II inhibitors during S and G2 phases points to a potential role for the WRN protein in a recombination pathway of double strand break repair in cooperation with topoisomerases I and II in the maintenance of genomic integrity.

The preferential sensitivity of WS cells to 4NQO and DNA cross-linking agents such as mitocmycin C suggests that WRN protein participates in the repair of only specific types of DNA lesions. The possibility that the defects in processing bulky adducts and DNA -links in WS cells are due to secondary effects of the mutator phenotype cannot be entirely ruled out. Hence, examination of the direct role of the WRN protein is warranted for its critical evaluation in the repair of DNA cross-links and bulky adducts induced by mitomycin C and 4NQO, respectively. Noticeably, the identification of the DNA substrates for the WRN helicase as well as its associated exonuclease activity revealed by recent studies have indicated interesting possibilities for the involvement of the WRN protein in diverse DNA metabolic activities.

An involvement of the WRN protein in basal transcription has recently been suggested.[39,40] It is unclear whether WRN helicase or exonculease activity is required for its participation in basal transcription. Shiratori et al[40] have showed that the WRN protein can be part of the RNA polymerase I associated complex. RNA polymerase I transcription is stimulated by WRN. The redistribution of WRN from the nucleus to the nucleolus and vice-versa upon transcriptional stimulation have led to the speculation that it may modulate the basal transcription in both nuclear and nucleolar environments. The extreme C-terminal region of WRN contains a nuclear localization signal that can direct a GFP-tagged (Green Flourescent Protein) WRN protein into the nucleus of human cells.[41] In addition, Suzuki et al[42] have identified two positively charged amino acids (arginine 1403 and lysine 1404) at the C-terminus of the WRN protein as a putative nucleolar localization signal. It is still unknown whether the primary biological function of WRN is in the nucleus or in nucleolus. The nucleolar distribution of WRN in human cells substantiates the assumption that WRN could be critical for nucleolar architecture.

Mode of Action of the WRN Helicase/Exonuclease in Pathways of Genomic Stability

Many signs of accelerated aging and the high probability of cancer characterize WS patients. These two features indicate that the *WRN* gene has a potential role in the maintenance of genomic stability. However, the underlying mechanism is far from clear. Although the WRN enzyme may participate in diverse DNA metabolic activities, the most compelling one is its presumptive role in DNA replication. One of the foremost phenotypic defects observed is their prolonged S phase coupled with a retarded proliferation efficiency. A reduced replication initiation sites and a slower replication fork migration were reported in WS cells due to a major defect in the replication machinery.[43-45] Importantly, the involvement of the WRN protein in the organisation of replication initiation complex is supported by studies on the WRN homologue in *Xenopus laevis*, FFA-1 (Focus Forming Activity 1). FFA-1 is responsible for replication foci formation.[46] FFA-1 precedes the recruitment of RPA to chromatin and its depletion affects RPA foci formation suggestive of its involvement in replication initiation. Interestingly, the focal distribution of Sgs1 (the WRN homologue in yeast) also coincides with the origin of replication complex (ORC). This complex is comprised of six proteins that are involved in the initiation of replication. However, a direct evidence for the role of Sgs1 in DNA replication comes from the study of Lee et al.[47] When a temperature sensitive mutation in Sgs1 is present in the background of another helicase Srs2 mutant, the cells are unable to replicate DNA at the restrictive temperature. Cells bearing either one of the mutations have the potency to replicate to the wild type level. Ribosomal DNA transcription by RNA polymerase I is also diminished in the double mutant suggestive of a dual role for Sgs1 and Srs2 in both replication and transcription. Analogous to Sgs1, WRN interacts with replication proteins such as RPA, polymerase δ and PCNA.[30,48,49] PCNA, a processivity factor for DNA polymerases δ and ε, is involved in replication and in diverse DNA repair pathways.[50] WRN interacts with PCNA through its N-terminal region. WRN also stimulates DNA polymerase δ activity in yeast independently of its helicase activity. This stimulation is lost in the absence of one of the three subunits of the DNA polymerase δ, pol 32p in *S. cerevisiae* which corresponds to the p66 subunit in the human DNA polymerase δ.[48]

In view of the demonstrated ability of Sgs1 and WRN helicases in unwinding the four way DNA junctions,[32] it is tempting to speculate that the WRN protein may play an essential role in resolving secondary structures thereby facilitating the reinitiation of DNA replication at stalled replication forks. The redistribution of the WRN protein from the nucleolus to foci containing RPA in the nucleoplasm in response to a replication blockage imposed by hydroxyurea supports this assumption. The exact functional significance of such foci formation remains obscure. The abolition of foci formation by aphidicolin, an inhibitor of DNA polymerases, indicates that active ongoing replication is a prerequisite for the formation of such foci. The formation of RPA specific foci after replication arrest does not require WRN as WS cells show efficient RPA foci formation. Nevertheless, the capacity of the WRN enzyme to unwind synthetic DNA secondary structures raise a strong probability of its involvement with RPA in replication reinitiation at stalled replication forks. An inability to reinitiate DNA replication may cause illegitimate recombinational events between daughter strands. Failure to resolve secondary DNA structures arising from the blockage of replication fork migration is considered to be the basis for the aberrant chromosome segregation and abnormal recombination events in $E.coli$[51] especially in the absence of the RecQ helicase. This may also be the case in WRN defecient cells. The interaction between WRN, PCNA and RPA suggests that WRN may play a role in the replicon initiation as well as in the elongation of DNA polymerases. The retardation of DNA synthesis elongation by DNA polymerases may be a cause for the prolonged S phase observed in WS cells. Given the possibility of a role for WRN in replication, abnormal S phase regulation is not entirely an unexpected feature of WS cells. Consistent with this, increased apoptosis was observed in WS lymphoblastoid cells in S phase after treatment with the cross-linking agent, mitomycin C as described in the preceding section.

The sensitivity of WS cells to 4NQO, topoisomerase inhibitors and 8-methoxypsoralen (8MOP) indicates that WRN helicase/exonuclease may have an important role in diverse DNA repair pathways as these agents induce a wide spectrum of DNA lesions including base damage, bulky DNA adducts, inter-strand cross-links and strand breaks. It is unknown whether WRN participates in repair either through interaction with other proteins or through its helicase/exonuclease acivities or a combination of both. The functional assembly of WRN with the recombination protein Rad51 upon replication arrest indicates that it may be involved in recombinational repair pathway.[38] Recent studies have shown that Ku70 and 80 proteins directly interact with the N-terminus region of WRN[52] and stimulate its exonuclease activity.[52-54] Ku70 and 80 heterodimer is involved in the nonhomologous end joining pathway of double strand break repair and Ku70/80 knock out mice are extremely sensitive to ionizing radiation. The direct interaction and stimulation of WRN exonuclease by Ku proteins suggest a role for WRN in double strand break repair activity. Blockage of WRN exonuclease in the degradation of DNA containing base modifications such as 8-oxoadenine and 8-oxoguanine is relieved by Ku proteins which not only stimulate WRN exonuclease activity but also enables the effective degradation of DNA containing these lesions.[55] Similar to WS cells, premature replicative senescence has been observed in Ku80 deficient cells. Ku80 deficient cells are rescued from senescence by reduced expression of the p53 protein.[56] The premature replicative senescence observed in WRN and ku80 deficient cells indicate that both WRN and Ku proteins may participate in a common pathway to prevent the onset of premature senescence. Additionally, WRN has been shown to be a substrate for phosphorylation by DNA-dependent protein kinase (DNA-PK) both in vivo and in vitro.[57,58] Although WS cells are not sensitive to double strand break inducing agents like X-rays and bleomycin, the observed increase in nonhomologous recombination in WS cells tends to suggest some kind of deficiency in the error prone ligation reaction during double strand break repair leading to deletion of large genomic regions.[59] A complex involving Mre11, Rad50 and the gene product of Nijmegen breakage syndrome has been shown to play important roles in double strand break repair and meiotic recombination in eukaryotes. Unlike the WRN exonuclease that cleaves 5' protruding strand, the Mre11 protein exhibits nuclease activity that cleaves 3' protruding strand at the junction of single and double stranded DNA. It is likely that both WRN and Mre11 work in coordination in cleaving the 5' and 3' protruding ends during double strand break repair.

A role for WRN in the base excision repair (BER) pathway is raised by the stimulation of Fen1 (Flap endonuclease 1) activity by WRN.[60] Fen1 is a structure specific endonuclease that removes the last primer ribonucleotide on the lagging strand. It cleaves a 5' flap that may result from strand displacement during either replication or BER reaction. The sensitivity of WS cells to 4NQO and 8MOP may be partially due to the inefficient removal of certain oxidative base lesions by long patch BER. This notion is supported by the WRN interaction with PCNA, which is yet another component of the long patch BER. Apart from its role in replication and repair, Fen1 seems to control recombination between short DNA sequences. Consistent with this, enhanced recombination between short DNA sequences has been observed in yeast Rad27/Fen1 mutants.[61] Hence, there is the possibility that WRN acts in a similar manner in association with Fen1.

The increased apoptosis observed in WS lymphoblastoid cells upon treatment with agents that cause DNA inter-strand cross-links suggests again an involvement of WRN in recombinational repair pathway. In addition, WS cells show increased apoptosis upon treatment with hydroxyurea and camptothecin both of which induce DNA double strand breaks through blockage of replication forks as discussed above. An homologous recombination repair pathway seems to play a critical role in the repair of either stalled or collapsed replication forks to permit reinitiation of replication.[61,62] Since WS cells are sensitive to both hydroxyurea and camptothecin, this suggests that WRN may play a vital role either in the repair of stalled replication forks or in the reinitiation of replication. Furthermore, the colocalization of WRN with the recombination protein Rad51 following hydroxyurea and camptothecin treatment indicates a functional interaction between these two proteins in the resolution of stalled replication forks.[38] Although the spontaneous level of Rad51 in the absence of damage is high, Rad51 foci formation in response to camptothecin treatment is reduced in WS cells.[63] Consistent with the spontaneous increase in Rad51 level, WS cells also display elevated level of DNA strand breaks detected by single cell gel electrophoresis assay.[63] Although WS cells resume DNA replication after damage by hydroxyurea and camptothecin, increased apoptosis and chromosomal damage observed in WS cells in S and G2 phases indicate that the replication recovery may be affected in the absence of the WRN protein.

The notion that RecQ helicases play a role in S phase checkpoint regulation has received a great deal of attention in recent years.[45] In yeast, Srs2 null mutant cells harbouring a temperature sensitive Sgs1 mutation are unable to replicate at restrictive temperatures although cells bearing either of the two mutations alone can replicate[47] as mentioned above. This indicates that both Srs2 and Sgs1 helicases have redundant yet essential functions in replication. In addition to replication, RecQ helicases can sense the damage that causes either replication fork collapse or impede DNA polymerase progression. The recruitment of WRN and BLM proteins to the sites of replication blockage seems to suggest such a possibility. Evidence from *E.coli* suggests that the stalled replication forks are unstable and the strand breakage often occurs due to fragility of unpaired DNA stretches[51] leading to strand invasion events. Sgs1 may provide through its interaction with topoisomerases the enzymatic machinery required to resolve the strand invasion events and for replication reinitiation from the stalled sites. Colocalization of WRN with RPA and Rad51 upon DNA damage and its unwinding activity toward synthetic replication fork substrates substantiate for such a role of WRN in DNA replication.

Finally, a role for WRN in genome surveillance is also raised by its interaction with the p53 tumor suppressor protein,[64] which plays important roles in cell cycle regulation and apoptosis in response to DNA damage. It is interesting to note that loss of WRN and p53 results in opposite phenotypic effect at the cellular level: loss of p53 delays the onset of cellular senescence while the loss of WRN accelerates senescence in WS cells. Purified WRN protein enhances RNA polymerase II transcription in vitro[39] and therefore it is reasonable to assume that WRN potentiates the transcription of p53 dependent genes after DNA damage enabling the coordinated regulation of cell cycle and repair events. The attenuation of p53 dependent apoptosis in WS cells tends to suggest a functional interaction between p53 and WRN.[65] The increased cancer incidence and genomic instability may be due to deregulation of apoptosis in the absence of functional WRN gene. WS cells show a greatly attenuated p53 induction after UV irradiation, a feature also noticed in radiation sensitive AT cells. The gene mutated in AT patients, *ATM*, encodes a serine/threonine kinase belonging to the family of phosphoinositol kinase family. The similarity observed between AT and WS cells in terms of p53 induction illustrate the requirement of both genes for efficient p53 induction after DNA damage. The lack of p53 induction observed in both WS and AT cells suggests that WRN may potentiate p53 induction in response to damage through either direct or indirect interaction with ATM kinase. Corroborating with this, a synthetic WRN peptide was shown to be a substrate for the ATM kinase.[66] The *ATM* gene functions upstream of p53 in both G_1 and G_2 phases of the cell cycle and loss of either p53 or ATM results in a failure in triggering DNA damage signalling pathways. Sgs1 acts upstream of yeast Rad53p, the yeast homologue of DNA damage dependent human Chk2/Cds1checkpoint kinase.[67] Both Chk1 and Chk2 kinases are regulated by yeast Mec1/Rad3, that are homologues of ATM. WRN may participate in a damage-signalling pathway through its interaction with ATM, ATM related kinase ATR and checkpoint kinases hChk1 and hChk2, all of which play a role in p53 induction following different types of DNA damage in mammalian cells. Although the interaction between WRN and ATM remains to be established, such a possibility is strongly indicated by certain phenotypic characteristics of both AT and WRN cells such as increased chromosomal instability, replicative senescence and accelerated telomere shortening. Noticeably, Sgs1 has recently been demonstrated to play a role in the telomere maintenance in telomerase negative yeast cells in an homologous recombination mediated manner.[68,69] A speculative model for the involvement of WRN in cell cycle regulation and DNA repair is shown in Figure 2. According to this model, replication dependent induction of DNA strand breaks or stalled replication forks may activate kinases like ATM, ATR and DNA-PK which in turn impose a S phase checkpoint control through phosphorylation of cell cycle specific kinases. Recruitment of RecQ helicases to stalled replication forks could be achieved through phosphorylation by ATM, ATR and DNA-PK as well. WRN and BLM helicases, through their interaction with topoisomerases, may then help to restore replication by resolving aberrant DNA structures at stalled replication forks.

Conclusions

The involvement of WRN in the regulation of S phase after DNA damage has been supported by a number of experimental evidences: (1) prolongation of S phase in WS cells, (2) sensitivity of WS cells to agents that cause replication dependent DNA strand breaks, (3) increased S phase apoptosis after treatment with 4NQO, camptothecin, hydroxyurea and DNA inter-strand cross-linking agents, (4) ability of the WRN helicase in the resolution of secondary DNA structures that are reminiscent of repair, replication and recombination intermediates, (5) ability of the WRN exonuclease in the cleavage of 5' protruding strand at the junction of single and double stranded DNA, (6) interaction of WRN with Fen1, PCNA and RPA that are involved in DNA replication and (7) suppression of illegitimate recombination between stalled replication forks probably through interaction with Rad51. Thus, it is clear from the foregoing account that WRN helicase/exonuclease may perform a critical role in the regulation of S phase and the restoration of replication initiation after DNA damage. The question of how to relate all the diverse pathological features of WS patients to a single mutated gene is partially answered by recent studies implicating a role for WRN in every aspect of DNA mediated transactions such as repair, replication, transcription and recombination as depicted in Figure 3. It remains to be established whether the multiple defects observed in WS patients are primary or secondary effects of *WRN* gene mutation.

Sensitivity of Werner's Syndrome Cells to DNA Damaging Agents

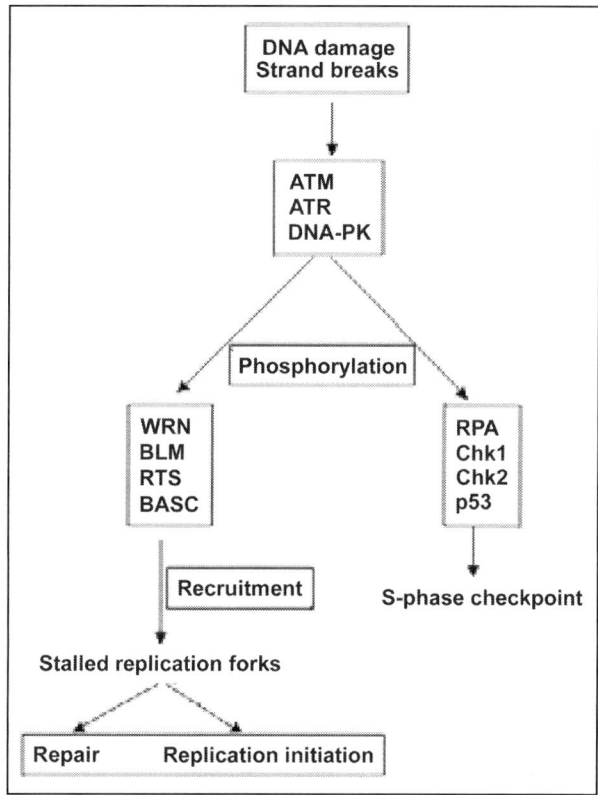

Figure 2. Potential involvement of WRN in cell cycle regulation and DNA repair.

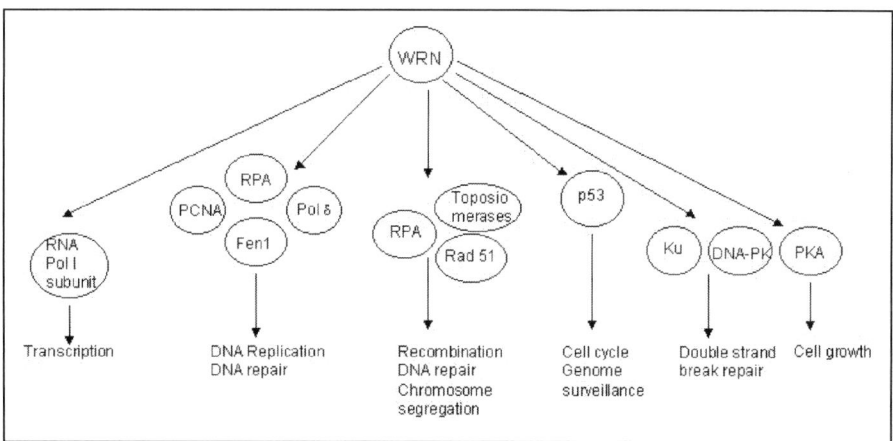

Figure 3. Predicted biological functions of the WRN helicase/exonuclease.

The relatively delayed onset of aging symptoms in WS patients indicates that the loss of the *WRN* gene is somehow compensated to certain extent by other helicases with similar functions. Identification of five RecQ-type helicases in humans as compared to one in *E.coli* indicates that different helicases with specialized functions are required to maintain the integrity of the complex human genome. The elevated sister chromatid exchanges in BS cells and the variegated translocation mosaicism in WS cells indicate that the loss of one helicase cannot always be compensated by another one in human. In addition, the specialized functions of each helicase may be critical for certain tissues and the loss of which may lead to tissue specific genomic instability leading to tumorigenesis. Another reason for the multi-system impairment observed in WS patients may be due to the disruption of WRN mediated protein interactions, which may be required for the maintenance of genomic stability. For example, Rad51 foci formation in response to replication blockage is defective in WS cells as discussed earlier. This may lead to an abnormal replication recovery, an increase in strand breaks and finally apoptosis.

Studies over the last several years have yielded valuable information regarding the biological consequences of human helicase mutants. It is quite unclear at the moment how these different RecQ helicases work in harmony to achieve genomic stability. Future research in this fascinating field will resolve the functional complexity of known and unidentified helicases associated with human diseases.

Acknowledgments

We dedicate this article to our mentor, Professor A.T. Natarajan, State University of Leiden, The Netherlands, on the occassion of his 75th birthday.

References

1. Balajee AS, Bohr VA. Genomic heterogeneity of nucleotide excision repair. Gene 2000; 250:15-30.
2. Grigorova M, Balajee AS, Natarajan AT. Spontaneous and X-ray-induced chromosomal aberrations in Werner syndrome cells detected by FISH using chromosome-specific painting probes. Mutagenesis 2000; 15:303-310.
3. Fukuchi K, Martin GM, Monnat Jr RJ. Mutator phenotype of Werner syndrome is characterized by extensive deletions. Proc Natl Acad Sci USA 1989; 86:5893-5897.
4. Gebhart E, Bauer R, Raub U et al. Spontaneous and induced chromosomal instability in Werner syndrome. Hum Genet 1988; 80:135-139.
5. Ogburn CE, Oshima J, Poot M et al. An apoptosis-inducing genotoxin differentiates heterozygotic carriers for Werner helicase mutations from wild-type and homozygous mutants. Hum Genet 1997; 101:121-125.
6. Prince PR, Ogburn CE, Moser MJ et al. Cell fusion corrects the 4-nitroquinoline 1-oxide sensitivity of Werner syndrome fibroblast cell lines. Hum Genet 1999; 105:132-138.
7. Hisama FM, Chen YH, Meyn MS et al. WRN or telomerase constructs reverse 4-nitroquinoline 1-oxide sensitivity in transformed Werner syndrome fibroblasts. Cancer Res 2000; 60:2372-2376.
8. Orren DK, Brosh Jr RM, Nehlin JO et al. Enzymatic and DNA binding properties of purified WRN protein: high affinity binding to single-stranded DNA but not to DNA damage induced by 4NQO. Nucleic Acids Res 1999; 27:3557-3566.
9. Franchitto A, Proietti De Santis L, Pichierri P et al. Lack of effect of caffeine post-treatment on X-ray-induced chromosomal aberrations in Werner's syndrome lymphoblastoid cell lines: a preliminary report. Int J Radiat Biol 1999; 75:1349-1355.
10. Pichierri P, Franchitto, A Mosesso P et al. Werner's syndrome lymphoblastoid cells are hypersensitive to topoisomerase II inhibitors in the G2 phase of the cell cycle. Mutat Res 2000; 459:123-133.
11. Pichierri P, Franchitto A, Mosesso P et al. Werner's syndrome cell lines are hypersensitive to camptothecin-induced chromosomal damage. Mutat Res 2000; 456:45-57.
12. Franchitto A, Pichierri P, Mosesso P et al. Catalytic inhibition of topoisomerase II in Werner's syndrome cell lines enhances chromosomal damage induced by X-rays in the G2 phase of the cell cycle. Int J Radiat Biol 2000; 76:913-922.

13. Lebel M, Leder P. A deletion within the murine Werner syndrome helicase induces sensitivity to inhibitors of topoisomerase and loss of cellular proliferative capacity. Proc Natl Acad Sci USA 1998; 95:13097-13102.
14. Gangloff S, McDonald JP, Bendixen C et al. The yeast type I topoisomerase Top3 interacts with Sgs1, a DNA helicase homolog: a potential eukaryotic reverse gyrase. Mol Cell Biol 1994; 14:8391-8398.
15. Watt PM, Louis EJ, Borts RH et al. Sgs1: A eukaryotic homolog of E. coli RecQ that interacts with topoisomerase II in vivo and is required for faithful chromosome segregation. Cell 1995; 81:253-260.
16. Ui A, Satoh Y, Onoda F et al. The N-terminal region of Sgs1, which interacts with Top3, is required for complementation of MMS sensitivity and suppression of hyper-recombination in sgs1 disruptants. Mol Genet Genomics 2001; 265:837-850.
17. Harmon FG, Kowalczykowski SC. RecQ helicase, in concert with RecA and SSB proteins, initiates and disrupts DNA recombination. Genes Dev 1998; 12:1134-1144.
18. Harmon FG, DiGate RJ, Kowalczykowski SC. RecQ helicase and topoisomerase III comprise a novel DNA strand passage function: a conserved mechanism for control of DNA recombination. Mol Cell 1999; 3:611-620.
19. Poot M, Yom JS, Whang SH et al. Werner syndrome cells are sensitive to DNA cross-linking drugs. Faseb J 2001; 15:1224-1226.
20. Yu CE, Oshima J, Fu YH et al. Positional cloning of the Werner's syndrome gene. Science 1996; 272:258-262.
21. Huang S, Li B, Gray MD et al. The premature ageing syndrome protein, WRN, is a 3'—>5' exonuclease. Nat Genet 1998; 20:114-116.
22. Shen JC, Gray MD, Oshima J et al. Werner syndrome protein. I DNA helicase and dna exonuclease reside on the same polypeptide. J Biol Chem 1998; 273:34139-34144.
23. Kamath-Loeb AS, Shen JC, Loeb LA et al. Werner syndrome protein. II Characterization of the integral 3' —> 5' DNA exonuclease. J Biol Chem 1998; 273:34145-34150.
24. Suzuki N, Shiratori M, Goto M et al. Werner syndrome helicase contains a 5'—>3' exonuclease activity that digests DNA and RNA strands in DNA/DNA and RNA/DNA duplexes dependent on unwinding. Nucleic Acids Res 1999; 27:2361-2368.
25. Opresko PL, Laine JP, Brosh Jr RM et al. Coordinate action of the helicase and 3' to 5' exonuclease of Werner syndrome protein. J Biol Chem 2001; 276:44677-44687.
26. Xue Y, Ratcliff GC, Wang H et al. A minimal exonuclease domain of WRN forms a hexamer on DNA and possesses both 3'- 5' exonuclease and 5'-protruding strand endonuclease activities. Biochemistry 2002; 41:2901-2912.
27. Gray MD, Shen JC, Kamath-Loeb AS et al. The Werner syndrome protein is a DNA helicase. Nat Genet 1997; 17:100-103.
28. Suzuki N, Shimamoto A, Imamura O et al. DNA helicase activity in Werner's syndrome gene product synthesized in a baculovirus system. Nucleic Acids Res 1997; 25:2973-2978.
29. Shen JC, Gray MD, Oshima J et al. Characterization of Werner syndrome protein DNA helicase activity: directionality, substrate dependence and stimulation by replication protein A. Nucleic Acids Res 1998; 26:2879-2885.
30. Brosh Jr RM, Orren DK, Nehlin JO et al. Functional and physical interaction between WRN helicase and human replication protein A. J Biol Chem 1999; 274:18341-18350.
31. Fry M, Loeb LA. Human werner syndrome DNA helicase unwinds tetrahelical structures of the fragile X syndrome repeat sequence d(CGG)n. J Biol Chem 1999; 274:12797-12802.
32. Constantinou A, Tarsounas M, Karow JK et al. Werner's syndrome protein (WRN) migrates Holliday junctions and colocalizes with RPA upon replication arrest. EMBO Rep 2000; 1:80-84.
33. Brosh Jr RM, Waheed J, Sommers JA. Biochemical characterization of the DNA substrate specificity of werner syndrome helicase. J Biol Chem 2002; 277:23236-23245.
34. Ohsugi I, Tokutake Y, Suzuki N et al. Telomere repeat DNA forms a large noncovalent complex with unique cohesive properties which is dissociated by Werner syndrome DNA helicase in the presence of replication protein A. Nucleic Acids Res 2000; 28:3642-3648.
35. Sun H, Bennett RJ, Maizels N. The Saccharomyces cerevisiae Sgs1 helicase efficiently unwinds G-G paired DNAs. Nucleic Acids Res 1999; 27:1978-1984.

36. Brosh Jr RM, Majumdar A, Desai S et al. Unwinding of a DNA triple helix by the Werner and Bloom syndrome helicases. J Biol Chem 2001; 276:3024-3030.
37. Liu Z, Macias MJ, Bottomley MJ et al. The three-dimensional structure of the HRDC domain and implications for the Werner and Bloom syndrome proteins. Structure Fold Des 1999; 7:1557-1566.
38. Sakamoto S, Nishikawa K, Heo SJ et al. Werner helicase relocates into nuclear foci in response to DNA damaging agents and colocalizes with RPA and Rad51. Genes Cells 2001; 6:421-430.
39. Balajee AS, Machwe A, May A et al. The Werner syndrome protein is involved in RNA polymerase II transcription. Mol Biol Cell 1999; 10:2655-2668.
40. Shiratori M, Suzuki T, Itoh C et al. WRN helicase accelerates the transcription of ribosomal RNA as a component of an RNA polymerase I-associated complex. Oncogene 2002; 21:2447-2454.
41. Matsumoto T, Imamura O, Goto M et al. Characterization of the nuclear localization signal in the DNA helicase involved in Werner's syndrome. Int J Mol Med 1998; 1:71-76.
42. Suzuki T, Shiratori M, Furuichi Y et al. Diverged nuclear localization of Werner helicase in human and mouse cells. Oncogene 2001; 20:2551-2558.
43. Takeuchi F, Hanaoka F, Goto M et al. Altered frequency of initiation sites of DNA replication in Werner's syndrome cells. Hum Genet 1982; 60:365-368.
44. Takeuchi F, Hanaoka F, Goto M et al. Prolongation of S phase and whole cell cycle in Werner's syndrome fibroblasts. Exp Gerontol 1982; 17:473-480.
45. Frei C, Gasser SM. RecQ-like helicases: The DNA replication checkpoint connection. J Cell Sci 2000; 113:2641-2646.
46. Yan H, Chen C, Kobayashi R et al. Replication focus-forming activity 1 and the Werner syndrome gene product. Nat Genet 1998; 19:375-378.
47. Lee SK, Johnson RE, Yu SL et al. Requirement of yeast SGS1 and SRS2 genes for replication and transcription. Science 1999; 286:2339-2342.
48. Kamath-Loeb AS, Johansson E, Burgers PM et al. Functional interaction between the Werner Syndrome protein and DNA polymerase delta. Proc Natl Acad Sci USA 2000; 97:4603-4608.
49. Lebel M, Spillare EA, Harris CC et al. The Werner syndrome gene product copurifies with the DNA replication complex and interacts with PCNA and topoisomerase I. J Biol Chem 1999; 274:37795-37799.
50. Tsurimoto T. PCNA binding proteins. Front Biosci 1999; 4:D849-D858.
51. Cox MM, Goodman MF, Kreuzer KN et al. The importance of repairing stalled replication forks. Nature 2000; 404:37-41.
52. Cooper MP, Machwe A, Orren DK et al. Ku complex interacts with and stimulates the Werner protein. Genes Dev 2000; 14:907-912.
53. Li B, Comai L. Requirements for the nucleolytic processing of DNA ends by the Werner syndrome protein-Ku70/80 complex. J Biol Chem 2001; 276:9896-9902.
54. Li B, Comai L. Functional interaction between Ku and the werner syndrome protein in DNA end processing. J Biol Chem 2000; 275:28349-28352.
55. Orren DK, Machwe A, Karmakar P et al. A functional interaction of Ku with Werner exonuclease facilitates digestion of damaged DNA. Nucleic Acids Res 2001; 29:1926-1934.
56. Lim DS, Vogel H, Willerford DM et al. Analysis of ku80-mutant mice and cells with deficient levels of p53. Mol Cell Biol 2000; 20:3772-3780.
57. Yannone S., Roy S, Chan DW et al. Werner syndrome protein is regulated and phosphorylated by DNA-dependent protein kinase. J Biol Chem 2001; 276:38242-3828.
58. Karmakar P, Piotrowski J, Brosh Jr RM et al. Werner protein is a target of DNA-dependent protein kinase in vivo and in vitro, and its catalytic activities are regulated by phosphorylation. J Biol Chem 2002; 277:18291-18302.
59. Runger T, Bauer C, Dekant B et al. Hypermutable ligation of plasmid DNA ends in cells from patients with Werner syndrome. J Invest Dermatol 1994; 102:45-48.
60. Brosh Jr RM, von Kobbe C, Sommers JA et al. Werner syndrome protein interacts with human flap endonuclease 1 and stimulates its cleavage activity. Embo J 2001; 20:5791-5801.
61. Negritto MC, Qiu J, Ratay DO et al. Novel function of Rad27 (FEN-1) in restricting short-sequence recombination. Mol Cell Biol 2001; 21:2349-2358.
62. Haber JE. DNA recombination: The replication connection. Trends Biochem Sci 1999; 24:271-275.

63. Pichierri P, Franchitto A, Mosesso P et al. Werner's syndrome protein is required for correct recovery after replication arrest and DNA damage induced in S-phase of cell cycle. Mol Biol Cell 2001; 12:2412-2421.
64. Blander G, Kipnis J, Leal JF et al. Physical and functional interaction between p53 and the Werner's syndrome protein. J Biol Chem 1999; 274:29463-29469.
65. Spillare EA, Robles AI, Wang XW et al. p53-mediated apoptosis is attenuated in Werner syndrome cells. Genes Dev 1999; 13:1355-1360.
66. Kim ST, Lim DS, Canman CE et al. Substrate specificities and identification of putative substrates of ATM kinase family members. J Biol Chem 1999; 274:37538-37543.
67. Frei C, Gasser SM. The yeast Sgs1p helicase acts upstream of Rad53p in the DNA replication checkpoint and colocalizes with Rad53p in S-phase-specific foci. Genes Dev 2000; 14:81-96.
68. Johnson FB, Marciniak RA, McVey M et al. The Saccharomyces cerevisiae WRN homolog Sgs1p participates in telomere maintenance in cells lacking telomerase. Embo J 2001; 20:905-913.
69. Cohen H, Sinclair DA. Recombination-mediated lengthening of terminal telomeric repeats requires the Sgs1 DNA helicase. Proc Natl Acad Sci USA 2001; 98:3174-3179.

CHAPTER 6

Yeast RecQ Helicases:
Clues to DNA Repair, Genome Stability and Aging

Rozalyn M. Anderson and David A. Sinclair

Abstract

The budding yeast *Saccharomyces cerevisiae* has been used as model for a wide range of cellular processes, including those related to the RecQ-associated progeroid disease, Werner's syndrome (WS). Investigations of RecQ function in these lower eukaryotes have produced a large body of data that is directly relevant to the function of the Werner's syndrome helicase, WRN, and other human RecQ members such as BLM. The yeasts studies point to a role for eukaryotic RecQ helicases in a number of key activities associated with DNA repair and replication, including the resolution of aberrant DNA structures, recombinational DNA repair, suppression of illegitimate recombination and the intra-S phase DNA damage checkpoint. Biochemical studies demonstrate that the budding yeast RecQ helicase, Sgs1, resolves the same DNA substrates as the human RecQ enzymes, suggesting that they share a common function. Many of the proteins that interact with Sgs1 also interact with WRN or BLM, further highlighting the relevance of the yeast studies. Mutation of *SGS1* results in telomere defects and symptoms of premature aging, two features of WS. Like WS cells, *sgs1* mutants also have phenotypes that are not aging-related and most likely stem from defects in recombinational repair and DNA damage signaling.

Yeast RecQ Helicases: Clues to Genome Instability and Aging

The bacterial *Escherichia coli* RecQ DNA helicase is the founding member of a highly conserved family, with RecQ homologues and orthologues identified in every species examined so far. RecQ is an important component of the RecF pathway that suppresses illegitimate recombination and mediates DNA repair at stalled DNA replication forks.[1,2] The RecQ helicase family has attracted considerable scientific attention since the discovery that at least three human diseases result from defects in some family members. Bloom's syndrome (BS), Werner's syndrome (WS) and a subset of cases of Rothmund-Thomson's syndrome are caused by defects in the RecQ-like genes *BLM*, *WRN* and *RECQ4*, respectively.[3] All three of these diseases lead to an increased incidence of particular cancers and the latter two diseases display some symptoms that resemble premature aging.

Although the precise cellular role of the eukaryotic RecQ helicases is not yet known, a picture is emerging from the study of simple eukaryotes such as budding and fission yeast. Over the past decade, these organisms have pointed to roles for RecQ helicases in a variety of cellular processes including DNA repair and replication, DNA damage checkpoints, telomere

---

Molecular Mechanisms of Werner's Syndrome, edited by Michel Lebel. ©2004 Eurekah.com and Kluwer Academic / Plenum Publishers.

maintenance and even the yeast aging process. In each of these processes the role of RecQ may be understood in terms of its ability to resolve aberrant DNA structures. There is only one RecQ homologue in the budding yeast *Saccharomyces cerevisiae*, *SGS1*, and the mutant phenotype has been characterized in some detail. There are clear parallels between the yeast Sgs1 and human RecQ proteins, WRN and BLM, both in terms of mutant phenotypes, interactions with other proteins and cellular function.

Phenotypes of RecQ Mutants

Elucidation of the role of any given protein is greatly facilitated by analysis of the mutant phenotype (Table 1). *Saccharomyces cerevisiae sgs1* phenotypes include an increase in homologous recombination,[4-7] homeologous (nearly identical) recombination and gross chromosomal recombination (GCR) events.[8] *SGS1* deficient cells also display a defect in the intra-S phase DNA damage checkpoint,[9] and are hypersensitive to a number of DNA damaging agents, including hydroxyurea,[9,10] H_2O_2,[11] UV (in stationary phase only),[11,12] methylmethane sulfonate (MMS)[10,13] and the topoisomerase inhibitors such as mitoxantrone and camptothecin.[14,15] In addition, *sgs1* cells have compounding defects in telomere maintenance in the absence of telomerase,[16-18] display chromosome segregation defects,[5] meiotic defects[6,19] and premature aging.[20]

There are a number of parallels between *sgs1* phenotypes and the phenotypes of WS and BS. A hyper-recombination defect has long been recognized as a feature of BS cells.[21,22] WS cells exhibit hyper-recombination between homeologous DNA sequences, resulting in the accumulation of GCR such as translocations and deletions.[23-25] Thus, in terms of recombination defects, *sgs1* cells exhibit features resembling both BS and WS cells. The hyper-recombination phenotype of *sgs1* cells for both illegitimate and homologous recombination is suppressed by expression of BLM or WRN, indicating that the human helicases likely have a similar function to Sgs1.[7] Given that increased homeologous recombination is a WS phenotype, it will be interesting to see whether WRN can suppress the increased homeologous recombination in *sgs1* strains.

WRN deficient cells display defects in S phase[26] and sensitivity to DNA damaging agents such as camptothecin,[27,28] 4-nitroquinoline 1-oxide (4-NQO)[29] and hydroxyurea.[26] Like *sgs1* strains, WS fibroblasts also show telomere defects in the absence of telomerase.[30-32] It is not clear whether WRN is involved in meiosis, but unlike BS patients, WS individuals can engender offsprings.[33] BLM localizes to meiotic chromosomes where it interacts with RPA (replication protein A).[34]

The premature aging phenotype of WS has been well documented.[35,36] Patients display a number of characteristics of aging and their disposition is thought to mimic old age. Similarly in yeast, mutations in *SGS1* result in premature aging phenotypes that stem from genome instability and defects in DNA repair.[37]

Like *sgs1* cells, the fission yeast *Schizosaccharomyces pombe rqh1* mutants are defective in the S phase checkpoint and are hypersensitive to hydroxyurea and UV (Table 2),[38-40] the latter being a hallmark of BS cells.[21,22] Mutant phenotypes in other fungal species point to additional roles for fungal RecQ helicases in PI 3-kinase-mediated DNA damage responses,[41] and post-transcriptional gene silencing.[42] It is striking that mutation of a single gene can give rise to such an array of phenotypes. Genetic evidence points to a role for yeast eukaryotic RecQ helicases in the detection and processing of damaged DNA or aberrant secondary structures during DNA replication and meiosis. Indeed all of the phenotypes described above can be at least partially understood in terms of defects in the resolution of aberrant DNA structures.

Table 1. Comparison of yeast and mammalian RecQ biochemical properties and mutant phenotypes

Properties/Mutant Phenotypes	Sgs1	WRN	BLM
Helicase activity	Yes	Yes	Yes
Exonuclease activity	No	Yes	No
Phosphorylation	Yes	Yes	Yes
Localization in nucleolus	Yes	Yes	Yes[a]
Localization in nuclear foci	Yes	Yes[b]	Yes
Hyper-recombination	Yes	Yes	Yes
Increased homologous recombination	Yes	No	Yes
Increased homeologous recombination	Yes	Yes	No
Gross chromosomal deletions	Yes	Yes	No
Premature aging phenotypes	Yes	Yes	No
Meiotic defect	Yes	No	Yes
Cell cycle checkpoint defect	Yes	Partial	Yes
Telomere maintenance defects[c]	Yes	Yes	No
UV sensitivity	No	No	Yes
MMS[d] sensitivity	Yes	Yes	Yes
Camptothecin[e] sensitivity	Yes	Yes	Yes

[a] In S-phase; [b] In response to DNA damage; [c] In absence of telomerase activity; [d] MMS= methy methane sulfonate, an alkylating agent; [e] A type I topoisomerase inhibitor.

Structures, Substrates and Localization

All RecQ helicases contain a central helicase domain, comprising 300-400 amino acids with seven conserved motifs. These include the Walker A and B boxes required for ATP-binding and a DEXH box that defines this family of helicases (Fig. 1).[4,5,43] The RecQ family has been divided into two classes based on whether or not they possess additional domains in their C- and N-termini. Class I RecQ helicases have additional domains at both the N- and C-termini, ranging in size from 1300-1500 amino acids.[44] Members of this class include the human helicases WRN, BLM and RECQ4, Sgs1 from the budding yeast *Saccharomyces cerevisiae*, Rqh1 from the fission yeast *Schizosaccharomyces pombe*, *qde-3* from the filamentous fungus *Neurospora crassa* and *musN* from *Aspergillus nidulans* (Table 2). Class II RecQ helicases include *E. coli* RecQ, human RecQL and RecQ5, and ORQA from *A. nidulans*, which are only ~500 amino acids in length and comprise little more than the helicase domain. None of the lower eukaryotic RecQ helicases possesses a WRN-like N-terminal exonuclease domain (see Fig. 1).[45-48] This may place limits on the utility of yeast as a WRN model. However, it is possible that Sgs1 recruits an exonuclease activity from another protein.

To date, Sgs1 from *S. cerevisiae* is the only lower eukaryotic RecQ helicase that has been studied at the biochemical level. Like RecQ helicases from mammals and *E. coli*, Sgs1 readily unwinds standard double-stranded DNA (dsDNA) substrates in the 3' to 5' direction in an ATP-dependent manner.[49,50] In addition to standard substrates, Sgs1 preferentially resolves other non-standard DNA substrates such as forked DNA, hairpins, four-way (synthetic Holliday) junctions and G-quadruplex (G4) DNA.[50-52] Both WRN and BLM can also resolve non-standard DNA substrates, including synthetic Holliday junctions and G4 DNA.[47,53-56] These findings indicate that a function of RecQ helicases may be to resolve abnormal and detrimental DNA secondary structures or recombination intermediates.

Table 2. Properties of fungal RecQ helicases

Organism	Name	Size (Amino Acids)	Substrates	Key Phenotypes of Mutant	Genetic Interactors	Physical Interactors	Putative Functions
Saccharomyces cerevisiae	Sgs1	1447	ssDNA, dsDNA, forked DNA, G-DNA, 4-way junctions	Hydroxyurea and MMS sensitive, suppresses slow growth of top3 mutants. Synthetically lethal/slow growing with srs2 mutation	srs2, top2, top3, slx1, mms4/slx2, mus81/slx3, slx4-6, rad50, rad51, rad52, rad16, rad24, MMR genes, mgs1	Top1, Top2, Rad16, hRad51, Rad53?, Mgs1/yWHIP?	Restart of stalled DNA replication, telomere repair/maintenance, S-phase damage checkpoint, Pol I transcription, delays aging
Saccharomyces pombe	rqh1 (hus2, rad12)	1328	unknown	Hydroxyurea and UV sensitive	top3+	unknown	Posttranscriptional gene silencing
Neurospora crassa	qde-3	1955	unknown	Post transcriptional gene silencing (quelling)-deficient, camptothecin sensitive, suppresses lethality of top3 mutation	qde-1, qde-2	unknown	Post-transcriptional gene silencing, DNA repair
Aspergillus nidulans	MUSN	1534	unknown	Suppresses PI 3-kinase(UVSB)-mediated DNA damage response	uvsC, uvsB, musP, orqA	unknown	Acts downstream of UVSB in the pathway regulating septum formation in response to DNA damage.
Aspergillus nidulans	ORQA	548	unknown	Extra-copies partially suppress MMS sensitivity of musN227 mutation	musN	unknown	Human RecQL5 homologue

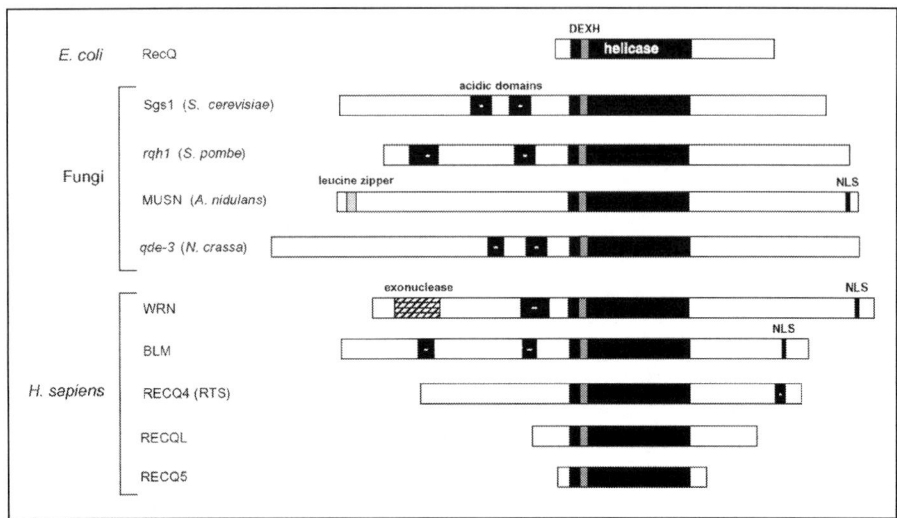

Figure 1. Structural comparison of yeast and human RecQ helicases. All RecQ helicases possess a central helicase domain (large black boxes) comprising 300-400 amino acids with seven conserved motifs including the Walker A and B boxes (required for ATP-binding) and a DEXH box that defines this family of helicases. Class I RecQ helicases have additional domains at both the N- and C-termini, ranging in size from 1300-1500 amino acids whereas Class II RecQ helicases comprise little more than the helicase domain. The N-terminal exonuclease domain of WRN and the leucine zipper domain of *Aspergillus nidulans* MUSN are unique for the family.

The fact that RecQ helicases are capable of unwinding G4 DNA as a substrate is of particular interest. G4 DNA is a highly stable structure that forms readily in vitro at G-rich single-stranded DNA (ssDNA).[57] Hoogsteen base-pairing between four guanines in a square planar configuration confers stability. G-rich sequences are found throughout eukaryotic genomes, particularly at telomeres and ribosomal DNA (rDNA) and these sequences may form G4 DNA during DNA repair, transcription or DNA replication. Although the existence of G4 DNA in vivo has not been demonstrated, there are a number of findings that indicate this DNA structure is biologically relevant. First, RecQ helicases are the only helicases known to resolve G4 DNA indicating that it is not a general property of helicases. Second, the yeast Hop1 protein binds robustly to G4 DNA and catalyzes its formation in vitro, and strains lacking Hop1 do not form mature synaptonemal complexes during meiosis.[58] Third, Sep1/Kem1 cleaves DNA in a G4 DNA-dependent fashion in vitro, and strains lacking this protein are defective in meiosis and undergo cellular senescence due to telomere shortening.[59-61] These genetic findings suggest that G4 DNA plays a crucial role in meiotic synapsis, telomere maintenance and DNA recombination.[58] This raises the possibility that some of the phenotypes of RecQ diseases result from an inability to resolve G4 DNA.

Structural analysis has shown that the BLM helicase can form oligomeric rings with four-fold and six-fold symmetry,[44] similar to other non-RecQ helicases such as the SV40 large T-antigen and *E. coli* RuvB.[62] The active form of WRN is thought to be a homotrimer.[45] Although it is not yet known if Sgs1 or other fungal RecQ helicases form oligomers, Brill and colleagues recently showed that Sgs1 and Top3 co-fractionate as a 1.3 MDa complex suggesting that Top3 and Sgs1 might exist as hexamers.[63]

Both WRN and BLM have nuclear localization signals (NLS) at their C-termini. Although Sgs1 lacks a consensus NLS, it is localized to the nucleolus[20] and in S phase it is localized to distinct foci that overlap with Rad53 at putative DNA replication forks.[9] WRN is also localized to the nucleolus under normal conditions.[64-66] Perhaps the G-rich composition of rDNA makes it particularly susceptible to G4 or other non-standard DNA structures whose resolution requires the activity of RecQ helicases.

Sgs1 is modified by phosphorylation, and while the biological basis for this modification is not yet known,[63] there are some clues from the human RecQ helicases. WRN protein is phosphorylated by DNA-PK in the presence of Ku and phosphorylation inhibits both catalytic activities of WRN.[67,68] In vivo, WRN is phosphorylated by DNA-PK in cells treated with DNA damaging agents[68] and phosphorylation regulates the relocalization of WRN from the nucleolus to a more diffuse pattern.[64] During mitosis BLM is re-localized from the matrix to the soluble fraction in the nucleus as a result of phosphorylation and phosphorylated BLM protein retains catalytic activity.[69] ATM and ATR are two PI 3-kinases that play a crucial role in DNA damage checkpoint response. In response to replication arrest, BLM is phosphorylated by the ATR kinase and this step is required for the localization of the RAD50-MRE11-NBS1 complex.[70] Irradiation also induces phosphorylation of BLM by ATM kinase and BLM is considered to be a downstream effector of the kinase.[71]

Physical Interactions

DNA Mismatch Repair (MMR) Proteins

The DNA mismatch repair (MMR) process is crucial for avoiding inappropriate recombination between similar but not identical sequences. Defects in MMR often lead to the accumulation of gross chromosomal rearrangements (GCR). In yeast, MMR is carried out by various mismatch repair protein complexes including Msh2-Msh3, Msh2-Msh6, and Mlh1-Pms1.[72] Studies of recombination between homeologous DNA sequences have identified a genetic link between eukaryotic RecQ helicases and DNA mismatch repair proteins. Yeast cells lacking *SGS1* or *MSH2* exhibit a 13- and 6-fold higher rate of homeologous recombination than wild type cells whereas cells lacking both genes exhibit a 96-fold higher rate.[8] A similar synergistic effect between *SGS1* and *MSH2* on the mutation rate and appearance of GCRs has also been observed.[73] These results indicate that Sgs1 and Msh2 function in independent pathways that suppress homeologous recombination or that they have redundant functions in the same pathway. In a recent study, Sgs1 has been shown to physically interact with Mlh1 (Fig. 2),[74] although the functional significance of this interaction is not yet clear.

The results in yeast indicate that a defect in the suppression of homeologous recombination may underlie the karyotypic abnormalities and genome instabilities in WS and possibly BS cells.[75-77] The redundancy between Sgs1 and MMR suggests that the BS and WS defects may be more severe in tissues where MMR is limiting, and this may explain the spectrum of sporadic tumors that characterize these diseases.[8] Also consistent with the yeast results, both WRN-mutant cells and *E. coli* RecQ mutants show an increased frequency of deletions between repeated sequences.[78,79]

In agreement with the genetic data from yeast, the mammalian BLM helicase has been shown to be part of a BRCA1-associated genome surveillance complex known as BASC.[80] The BASC complex includes MSH2, MSH6, MLH1 (mismatch repair proteins), ATM (ataxia telangiectasia mutated), BRCA1 (breast cancer protein-1), RMN complex RAD50-MRE11-NBS1 (telomere/DNA break repair complex), replication factor C and BLM. The Sgs1-Msh1 interaction appears to have been evolutionarily conserved because BLM physically interacts and co-localizes with hMLH1.[74] Interestingly, like the yeast and human RecQ helicases, Msh2, Msh6 and Mlh proteins bind to synthetic Holliday junctions, implying that they may be involved in recombinational repair of replication fork damage.[74,81,82]

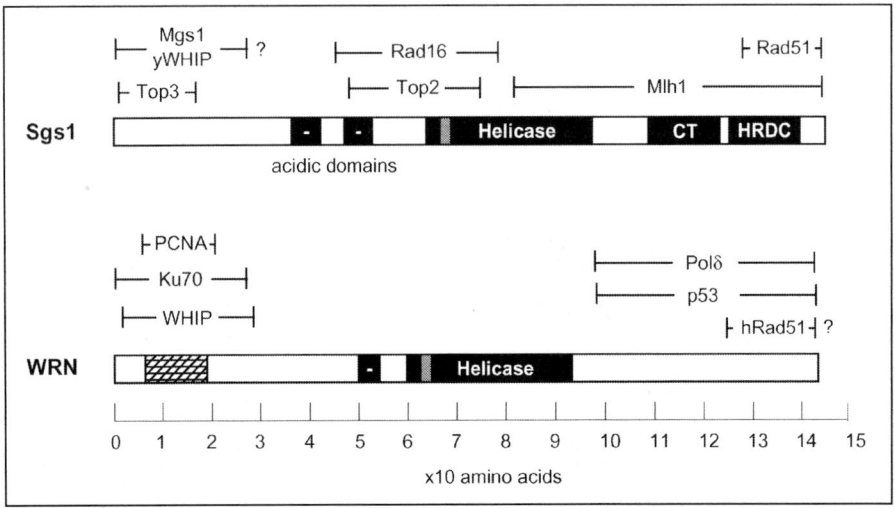

Figure 2. Domains of Sgs1 and WRN. Minimal domains for Sgs1 and WRN interactions are shown, as determined by co-immunoprecipitation and/or two-hybrid analyses. The C-termini of Sgs1 and BLM and possibly WRN, interact with Rad51. The N-terminus of WRN interacts with WHIP1 and genetic evidence suggests that the S. cerevisiae homologue of WHIP, Mgs1, interacts with Sgs1. Additional WRN interactors, whose minimal WRN-interaction domains have not been defined, include RPA (replication protein A), SUMO-1 (small ubiquitinating related modifier) and Ubc9 (SUMO-1-conjugating enzyme).

Mgs1, the Yeast WHIP Homologue

A two-hybrid screen for mouse WRN-interacting proteins identified a novel protein designated WHIP (for WS helicase-interacting protein).[83] WHIP shows partial homology to the replication factor C (RFC) family of DNA-dependent ATPases that load the PCNA/Pol30p clamp onto a primed DNA template.[15,84] Murine WHIP shares 45% homology with the Mgs1 (maintenance of genome stability 1) protein of S. cerevisiae.[83] Mgs1 possesses DNA-dependent ATPase and ssDNA annealing activity and may be required to achieve the correct DNA topology during events such as recombination.[15] Deletion of MGS1 exacerbates the short life span of sgs1 strains (~6 divisions) but partially relieves the MMS sensitivity of sgs1 mutants, demonstrating that MMS sensitivity and life span are separable. Strains lacking MGS1 grow normally and are not sensitive to MMS. However, the average life span of individual cells is ~75% that of wild type cells (~21 vs. 27 divisions).[15] It is likely that under normal growth conditions Mgs1 does not act in a pathway with Sgs1. However, in the presence of DNA damage, Mgs1 acts in the same pathway upstream of Sgs1.[15] This is consistent with the apparent dual role for Sgs1: to suppress recombination under normal conditions and to induce recombination under conditions of excessive DNA-damage.[12,85] An attractive hypothesis is that, in the presence of DNA damage, the DNA-strand separating activity of Sgs1 is coupled to the DNA annealing activity of Mgs1.[83]

Topoisomerases

There is now convincing evidence that RecQ helicases can act in concert with topoisomerases to disrupt joint DNA molecules. Topoisomerases alter the degree of DNA supercoiling during replication and transcription and resolve joint DNA molecules that arise during mitosis and meiosis.[86] There are two classes of DNA topoisomerases. Type I topoisomerases make ssDNA nicks and can relax supercoiled DNA, whereas Type II topoisomerases can make dsDNA breaks and pass one dsDNA molecule through another. In *S. cerevisiae*, four topoisomerases have been described, three of which are nuclear and one mitochondrial.[86] A number of findings strongly indicate that Sgs1 interacts with all three nuclear topoisomerases to achieve different tasks. The three nuclear enzymes, Topos I, II and III are encoded by the *TOP1, TOP2* and *TOP3* genes, respectively. Yeast Topo I (a type II enzyme) is required for DNA replication,[87] mitotic chromosome condensation[88] and general transcriptional repression in stationary phase.[89] Deletion of *TOP1* has no effect on growth rate or viability under normal growth conditions.[90,91] Topo II, another type II enzyme, is required for the resolution of intertwined chromosomes during mitosis and meiosis and is essential for viability in yeast.[92-94] Topo III is a type I enzyme that makes a transient covalent linkage to the 5'-end of the cleaved DNA[95] and recombinant Top3 only acts on negatively supercoiled ssDNA.[95] Yeast strains lacking the *TOP3* gene grow extremely poorly,[95,96] accumulate in G2/M, are sensitive to DNA damaging agents,[11,97] have severe meiotic defects[98,99] and exhibit hyper-recombination between repetitive sequences such as telomeres and rDNA.[87,96,98]

The *SGS1* gene, was originally identified as a loss-of-function mutation that suppressed the slow growth of topoisomerase III (*top3*) mutants and was cloned by virtue of its interaction with Top3 in a two-hybrid system.[4,6] The interaction between Sgs1 and Top3 is DNA independent,[63] raising the possibility that Sgs1 may be involved in recruiting Top3 to the DNA. Deletion of *SGS1* partially suppresses the hyper-recombination and meiotic defects of *top3* cells[99] but not their sensitivity to DNA damaging agents.[63] The suppression of slow growth in *sgs1 top3* cells is rescued by BLM but not by WRN, indicating that this interaction may be more relevant to BS than to WS.[100] The interaction between *SGS1* and the other two topoisomerases is less well understood. *SGS1* was also identified in a 2-hybrid screen for topoisomerase II (Top2) interactors,[5] suggesting a physical interaction between these two proteins. A genetic interaction has been observed between *SGS1* and *TOP1* with the double mutant displaying slow growth though the individual deletions had no effect on growth rate.[5]

A number of subsequent genetic and biochemical studies have defined the minimal Sgs1 domains required for its biochemical or genetic interaction with the three topoisomerases (Fig. 2). The Top2 interaction domain of Sgs1 lies between residues 466-746, but the functional significance of the Sgs1-Top2 interaction is not yet clear.[5,101] The first 100 amino acids are required for Sgs1 function in the absence of Top1[10,101] and this region also physically interacts with Top3.[63,101,102] In other words, the Sgs1-Top3 interaction domain and the helicase domain are sufficient to prevent the slow growth and hyper-recombination of *sgs1 top1* mutants.[101] These complex physical and genetic interactions are consistent with a model in which a major function of Sgs1 is to direct Top3 to a specific substrate that occurs more frequently in the absence of Top1.[101,103,104] RecQ has recently been shown to interact with and stimulate Topo IIIs from *E. coli* and *S. cerevisiae* to fully catenate dsDNA molecules.[105] Thus, in the absence of Sgs1, Top3 likely has a limited range of substrates but in combination with Sgs1, Top3 may act on a far wider range of DNA substrates including dsDNA.[102]

The interaction between RecQ helicases and type I topoisomerases appears to be well conserved. In *S. pombe*, cells lacking the *top3⁺* gene are defective in chromosome segregation and most cells die within the first five divisions.[73] This lethality is partially suppressed by deleting *rqh1⁺*. In humans, the BS helicase (BLM) physically interacts with one of the two Top3 isozymes, Top3α[106] and RecQ5 co-immunoprecipitates both Top3 isozymes.[72] Heterologously expressed hTop3β interacts with Sgs1 in yeast and partially rescues the slow growth defect of *top3* mutants[107] confirming conservation of function between the Top3 proteins. Interestingly, a chimeric protein comprised of Top3 fused to a truncated Sgs1 (lacking the first 106 amino acids) can complement the sensitivity of *sgs1* strains to DNA damaging agents, demonstrating that the two proteins likely form a functional unit.[103] In fact, in the bacterium *Sulfolobus acidocaldarius* the putative helicase and topoisomerase activities are present on the same polypeptide.[108]

The physiological relevance of the genetic interaction between *SGS1* and *TOP1* is a matter of speculation. However, there is some evidence that this interaction is conserved in humans. Homozygous WS embryonic stem (ES) cells from mice show defects in some DNA repair systems exhibiting higher mutation rates than wild type cells and increased sensitivity to topoisomerase inhibitors.[27] Human WS cells also show enhanced sensitivity to anti-topoisomerase drugs.[26,28] Furthermore, studies in both mouse and human demonstrate that topoisomerase I co-immunoprecipitates with WRN.[109]

Rad51

In eukaryotes, a central step in homologous recombination is catalyzed by Rad51,[110,111] the homologue of *E. coli* RecA.[1] Rad51 is thought to bind to ssDNA and align it with a homologous duplex to promote an extensive strand exchange between them.[110] There is now good evidence that mammalian RecQ helicases interact with Rad51 and that this association has been evolutionarily conserved. BLM and RAD51 interact directly via residues in the N- and C-terminal domains of BLM.[112] Consistent with this, BLM co-localizes with a subset of RAD51 nuclear foci following ionizing radiation damage[112,113] and during meiosis.[114] Although WRN and RAD51 have not been shown to interact physically, they have been observed to co-localize in DNA damaged cells.[115] WS cells also exhibit a defect in induction of RAD51 foci in response to DNA damage during S phase.[116] Like BLM, the C-terminal domain of Sgs1 (residues 1299–1447) interacts with yeast Rad51 in a two-hybrid assay.[112] There are also complex genetic interactions between *SGS1* and *RAD51* (see below). The evolutionary conservation of the interaction between RecQ helicases and Rad51 implies that these two classes of proteins together perform a fundamentally important role during DNA metabolism.

Rad16

S. cerevisiae RAD16 and *RAD7* genes function together in the nucleotide excision repair (NER) of transcriptionally inactive DNA.[117] Rad16 and Rad7 exist as a tight complex named NER factor 4 (NEF4) that is suspected to translocate on DNA in search of UV lesions.[118,119] A physical interaction between Rad16 and Sgs1 was identified using a two-hybrid screening approach.[11] N-terminally truncated Rad16 (residues 82-790) was shown to co-immunoprecipitate with a core Sgs1 domain (residues 421-792) when expressed at high levels. Genetic analyses have shown that there are synthetic effects between *SGS1* and *RAD16*. Deletion of *RAD16* in an *sgs1* strain partially suppresses MMS sensitivity whereas deletion of *SGS1* in a *rad16* strain greatly increases sensitivity to DNA damage by UV and 4-Nitroquinoline 1-oxide (4-NQO). These results imply that Sgs1 and Rad16 may function in parallel and partially redundant pathways to repair UV and 4-NQO damage.[11] Regarding MMS-mediated DNA damage, Rad16 may initiate a process that requires Sgs1 for its resolution or Rad16 may direct Sgs1 to initiate a lethal process.

Genetic Interactions

The ability to perform genetic analyses in yeast allows for the rapid identification of interacting genes in a manner that is difficult to perform in mammals. While often these studies provide clues about pathways that operate in parallel they also allow for the identification of genes that operate in the same pathway, i.e., the same epistatic group.

SLX1/4, HEX3/SLX8 and *MMS4/MUS81*

A screen for mutations that result in lethality when combined with an *sgs1* deletion (i.e., synthetically lethal mutations) resulted in the identification of six complementation groups.[120] There is good evidence that the *so-called SLX* gene pairs (*SLX1/4, HEX3/SLX8* and *MMS4/*

MUS81) encode heterodimeric complexes that act in pathways parallel to Sgs1-Top3.[120,121] None of the *SLX* genes is essential but all null mutations are lethal in combination with mutations in *TOP3*. Subsequent analyses have placed the six complementation groups into three pairs, each with its own particular set of phenotypes.

The *MMS4/MUS81* gene pair has been characterized to a greater extent than the other *SLX* genes. Mutations in either *MMS4* or *MUS81* result in weak UV sensitivity, poor growth in combination with a *top1* mutation, and a severe defect in sporulation (a process requiring meiosis).[120] The sporulation defect can be partially suppressed by deletion of *SPO11*, which is required for the formation of double-strand breaks (DSBs) prior to meiosis I. The *mms4/mus81* sporulation defect can be fully suppressed by deleting both *SPO11* and *SPO13*, the latter of which causes cells to bypass meiosis I, indicating that these two genes are required for the processing of recombination intermediates during meiosis. Similarly, the *S. pombe mus81*⁺ gene is synthetically lethal with mutations in *rqh1*⁺ and mus81 mutants fail to sporulate.[122] In *S. pombe* Mus81 was found to interact with the Cds1 checkpoint kinase in a 2-hybrid screen[122] and in *S. cerevisiae* Mms4 interacts with the meiotic checkpoint kinase Mek1,[123] indicating that this complex is intimately involved in checkpoint responses.

The *MMS4/MUS81* pair encodes proteins with homology to the yeast Rad1-10 endonuclease (mammalian XPF-ERCC1), which functions in nucleotide excision repair.[124] Recent work by Brill and colleagues showed that Mms4 and Mus81 form a dimer that can cleave Y-forms of DNA and has a preference for branched duplex DNA and synthetic replication fork substrates.[121] These and other findings imply that the Mms4-Mus81 complex is required during S phase to bypass stalled DNA replication forks, particularly in the absence of Sgs1-Top3, which may remove the nascent DNA strand. The Mms4-Mus81 endonuclease may cleave the leading strand template and the DNA break created by this cleavage may recombine with the sister chromatid through an unknown mechanism. In meiosis, Mms4-Mus81 may resolve recombination intermediates that occur during the initiation of meiotic recombination.[121] The role for Sgs1 in meiosis has been traced to the first meiotic division.[6] The *sgs1* null strain shows a reduction in the levels of meiotic recombination perhaps reflecting a processing difficulty in cells lacking the helicase. The meiotic defect can be complemented by expression of an allele of *SGS1* that lacks helicase activity.[19] The interaction described here may shed some light on pathways that operate in higher eukaryotes. There may be a mammalian endonuclease that operates in parallel to the exonuclease activity in WRN.

SRS2

The *S. cerevisiae SRS2/HPR5* gene encodes a 3' to 5' DNA helicase[125,126] with high homology to the bacterial DNA helicases UvrD and Rep. No homologue of *SRS2* has yet been found in mammals. Yeast *SRS2* shares several phenotypic features with *SGS1* including a sensitivity of mutants to DNA damaging agents such as MMS and UV and a short average lifespan.[14,126-129] The UV sensitivity of *srs2* cells is suppressed by mutations in genes controlling homologous recombination, such as *RAD51*, *RAD55* or *RAD57*. This points to a role for Srs2 in directing DNA damage into recombinational repair.[12,128,130-133] Srs2 is phosphorylated in response to intra-S phase DNA damage in a checkpoint dependent manner.[134] Furthermore, srs2 mutants fail to activate Rad53 properly. Genetic analysis places *SRS2* in a DNA damage pathway that operates in parallel to the checkpoint genes *RAD17/RAD24*.

The fact that overexpression of *SGS1* can complement the MMS and HU (hydroxyurea) sensitivity of *srs2* mutants implies that the functions of Sgs1 and Srs2 in DNA repair are similar.[135] Interestingly, *SGS1* overexpression does not increase the average life span of *srs2* mutants whereas maximum life span is significantly increased.[135] As discussed in further detail below, this finding is similar to that of McVey and colleagues who concluded that the life span of *sgs1* and *srs2* mutants is the product of two independent processes: stochastic death and aging.[129] The most likely explanation for the effect of *SGS1* overexpression on the life span of *srs2* strains is that additional Sgs1 delays aging but does not complement stochastic death.

Although strains lacking the single deletions of *SRS2* or *SGS1* grow normally, strains lacking both are inviable or grow extremely slowly, depending on the strain.[101,129,136] The severe phenotype of the double mutants appears to be the result of defective DNA polymerase I transcription and, in particular, an inability to complete DNA replication.[12,136] Doubly mutant cells usually arrest in the G2/M phase of the cell cycle with their DNA wedged in the bud neck. As discussed below, the poor growth of the *sgs1 srs2* double mutant is significantly improved by abolishing homologous recombination, indicating that the cell cycle arrest phenotype is due to an inability of cells to resolve recombination intermediates during DNA replication.[12]

The apparent absence of a mammalian homologue for Srs2 suggests that the Srs2-RecQ interaction may be peculiar to yeast. The only known homologues are in prokaryotes, so perhaps this interaction reflects a more ancient pathway that has evolved divergently in higher eukaryotes with other candidates fulfilling the role played by *SRS2*. Nonetheless, the genetic interaction between *SRS2* and *SGS1* raises the possibility that the mammalian RecQ helicases may operate in parallel with other helicases.

Other DNA Repair Genes

A number of genetic modifiers of the *sgs1 srs2* synthetic slow growth/lethality have been identified. *RAD50* and *MRE11* are involved in DNA double-strand break (DSB) repair by non-homologous DNA end-joining (NHEJ), gene conversion and ssDNA annealing (SSA).[137,138] Both *RAD50* and *MRE11* are essential for the viability of *sgs1 srs2* strains. Similarly, *RAD1*, a gene required for SSA,[139] is essential for viability of *sgs1 srs2* strains.[129] Deletion of *RAD59* or *MSH2*, two genes that affect the efficiency of SSA, also reduce the viability of *sgs1 srs2* mutants, though to a lesser extent than *RAD50*, *MRE11* and *RAD1* deletions. These results can be explained by a model in which *sgs1 srs2* cells are defective in homologous recombination and use the alternative SSA pathway of DNA repair. Those *sgs1 srs2* cells that attempt to proceed with DNA repair by homologous recombination have difficulty completing the process and can not reverse it.[129]

This model affirms genetic data showing that the viability of the *sgs1 srs2* double mutant is greatly improved by deleting genes involved in homologous recombination, namely *RAD51*, *RAD55* and *RAD57*.[12] The epistatic relationship suggests a role for Sgs1 in a later step of the recombination process that takes place after strand invasion.[12] Sgs1 may have a role in promoting the maturation of recombination intermediates, whereas Srs2p may reverse otherwise abortive recombination intermediates.[131,132] In the *sgs1 srs2* double mutant, recombination structures formed during replication may never be resolved, leading to activation of the intra-S phase checkpoint, cell cycle arrest and cell death.[12]

The Cellular Function of RecQ Helicases

S Phase DNA Damage Checkpoint

Eukaryotes possess a well-conserved system of checkpoints that monitor and elicit a response to DNA damage or aberrant DNA structures (Fig. 3).[140-142] These checkpoints are crucial for genome stability because they can delay DNA replication and cell cycle progression in G1 or G2 or S phase until DNA repair is completed. Although the mechanism is still not well understood, mammalian cells appear to delay S phase by slowing the progression of replication forks and delaying the firing of late replication origins. The ataxia telangiectasia mutated

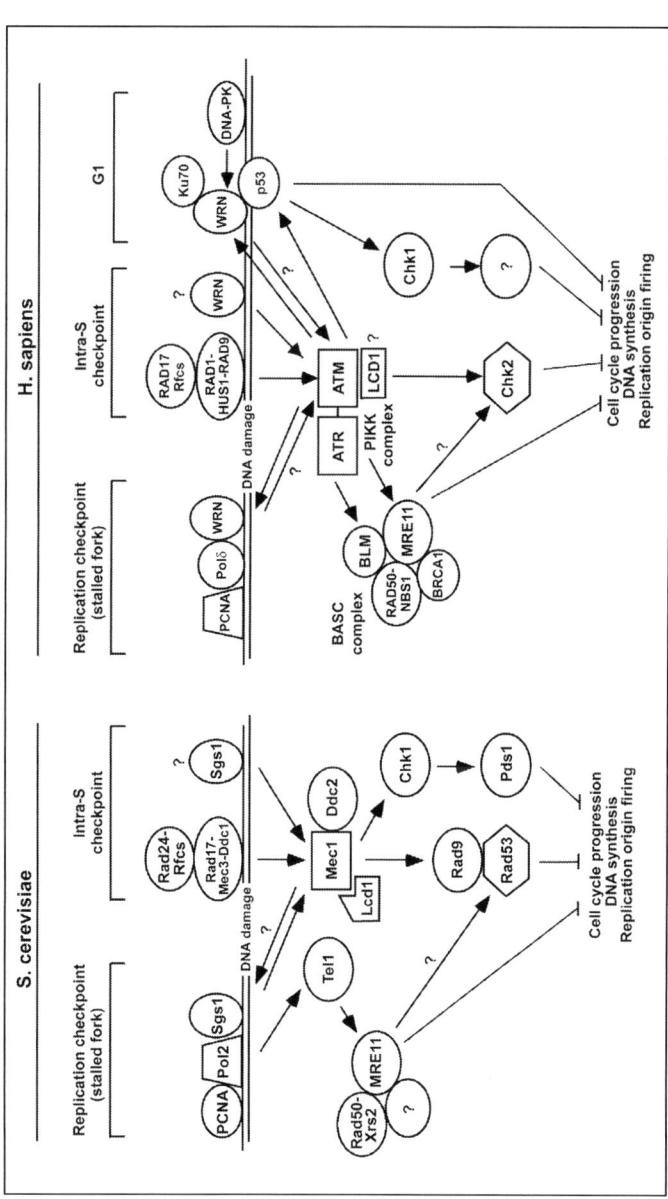

Figure 3. Schematic model of yeast Sgs1 and human WRN RecQ helicases in DNA damage detection and signaling. In yeast, the PI 3-kinase Mec1 is central to the DNA damage response and its homologue, Tel1 is partly redundant. Humans possess two Mec1 homologues, ATM and ATR, that form a complex known as PIKK. During S phase, Sgs1 and WRN associate with DNA damage (directly or after processing) and recruit Mec1/PIKK to the site of damage. The Mec1-associated protein Lcd1 and checkpoint clamp proteins Rad24/RAD17, Rad17/RAD1, Mec3/HUS1, Ddc1/RAD9 assist in the recruitment of Mec1/PIKK. Activated Mec1/PIKK phosphorylates the Rad53/Chk2 kinase, which act as effectors of the DNA damage signal. Genetic studies of the intra-S phase checkpoint place SGS1 upstream of TEL1, during G_1 or G_2. WRN associates with DNA damage where it is activated by DNA-PK and/or PIKK. WRN expression is controlled by p53 and WRN may recruit p53 to sites of damage to invoke a DNA damage response. S. cerevisiae has no p53 homologue and there is currently no evidence that Sgs1 participates in a G_1 or G_2 checkpoint response. Arrows indicate demonstrated or putative interactions based on genetic and biochemical evidence. Question marks represent predicted interactions or proteins based on homology between yeast and human pathways.

(ATM) PI 3-kinase[143,144] and the hCHK2/CDS1 kinase are key mediators of this checkpoint pathway.[145,146] In *S. cerevisiae*, two S phase signals can generate a checkpoint response. DNA damage triggers an intra-S phase checkpoint which signals via three main branches involving Chk1, Mre11 and Rad53, the yeast homologue of human CHK2.[147] Stalled DNA replication forks trigger a replication checkpoint, which requires the Dun1 kinase and some components of the DNA replication apparatus such as PCNA and DNA polymerase ε (Pol2). Both checkpoints require *MEC1*, a yeast ATM homologue whereas *TEL1*, another yeast ATM homologue, clearly plays a redundant role.[148-150]

Recent data points to a role for yeast RecQ helicases in the intra-S phase DNA damage checkpoint. During G1, Sgs1 is barely detectable by Western blotting. However, during S phase, levels of *SGS1* expression increase dramatically.[9,151] This increase in expression is due in part to the presence of two Swi4-Swi6 cell cycle box elements in the *SGS1* promoter.[151] Consistent with a role in DNA repair and checkpoint signaling, co-immunolocalization studies show that during S phase Sgs1 overlaps with Rad53 foci.[9] Moreover, strains lacking *SGS1* or *TOP3* are partially defective in the intra-S phase DNA-damage checkpoint and defective in Rad53 phosphorylation following DNA damage in S phase.[9,97] Analyses of the gross chromosomal rearrangements (GCR) in various mutants indicates that *SGS1* and *RAD24* are redundant in the S phase checkpoint and the *SGS1* branch acts predominantly through *TEL1*.[8,147] These studies also imply that Rad53 plays a relatively minor role in the *SGS1* intra-S phase pathway but a major role in the *RAD24* pathway.[147] Genetic analyses of the *S. pombe* response to UV- and γ radiation indicates that *rqh1+* acts upstream of an *S. cerevisiae DUN1* homologue, *rad9+*.[32] In a slightly different role than Sgs1, Rqh1 appears to regulate exit from the S phase checkpoint, rather then progression through S phase.[38,40] Together, the yeast studies indicate that a primary function of RecQ helicases is to detect DNA damage and problems during DNA replication and elicit a DNA damage response.

This idea is consistent with recent genetic and biochemical findings in mammals. Like Sgs1, *BLM* expression also is regulated during the cell cycle, accumulating to high levels in S phase and sharply declining by G1.[71] BS cells also display a partial defect in the γ irradiation-induced G2/M cell cycle checkpoint[71] and are highly sensitive to UVC irradiation during G1 to early S phase.[152] The mammalian ATM kinase acts upstream of the tumor suppressor p53 to prevent cell cycle progression and promote DNA repair and apoptosis.[153] Interestingly, both BLM and WRN functionally interact with the C-terminus of p53,[154-156] and p53-mediated apoptosis is attenuated in both BS and WS cells.[156,157] Cells lacking p53 have decreased accumulation of BLM in promyelocytic leukemia (PML) nuclear bodies,[157] the structural entities that recruit other cellular proteins such as Rb (retinoblastoma protein), SUMO-1 (small ubiquitin-related modifier-1) and UBC9 (a SUMO-conjugating enzyme).[158] Recently WRN has also been shown to interact with both SUMO-1 and UBC9.[83]

WRN is phosphorylated and thereby inactivated by DNA-PK, a kinase implicated in UV damage-induced arrest. DNA-PK is implicated in UV damage-induced replication arrest.[159] Genetic analysis suggests that Ku and Rad54 operate in parallel as competing pathways at sites of DNA damage in S phase.[24] Perhaps the DSB repair pathway requires the inactivation of WRN. WS and *sgs1* cells do not generally show UV sensitivity and do not show defects in DSB repair.[9] BLM is phosphorylated by ATM in response to irradiation-induced DNA damage[69] and is a downstream effector of the kinase. This fits with the recent finding that BLM is part of the BASC complex in which the BRCA1 tumor suppressor protein and ATM reside.[80] BLM is phosphorylated by the Rad3-related ATR kinase and this step is required to relocalize the RAD50-MRE11-NBS1 complex to nuclear structures following exposure of cells to HU.[70] This finding supports those made earlier in yeast indicating that RecQ helicases recognize and recruit repair factors to stalled DNA replication forks, then participate in the repair process.[9,160]

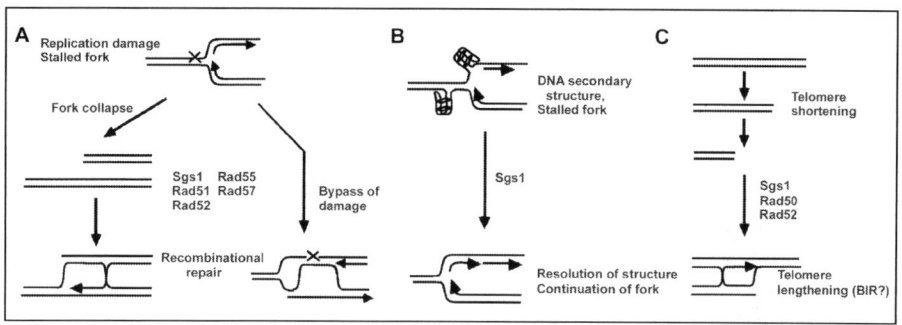

Figure 4. Model for Sgs1 function at stalled replication forks and telomeres. A) DNA damage or DNA secondary structures (indicated by cross) cause a DNA replication fork to stall. A stalled form is an inherently unstable structure that will collapse if ssDNA is nicked. Sgs1 mediates recombinational repair of the resulting DNA break by break-induced DNA replication (BIR) mediated by Sgs1 and Rad51 (a RecA homologue). DNA strand exchange and D-loop formation is facilitated followed by branch migration and Holliday junction resolution. B) Sgs1 may resolve DNA secondary structures such as G4-DNA directly, allowing the replication fork to proceed. C) In the absence of telomerase, cells with critically short telomeres can bypass senescence by activating a telomerase-independent replication-mediated pathway that may resemble BIR. Telomeres generated by this Rad50-dependent pathway are reminiscent of the very long "ALT" telomeres of some immortal mammalian cell lines and tumors.

DNA Replication and Repair

Eukaryotic RecQ helicases are implicated in numerous aspects of DNA metabolism including the resolution of aberrant DNA structures, suppressing illegitimate recombination and restarting DNA replication. A difficulty in studying these processes is that the distinction between them becomes blurred in the events surrounding replication. Is the resolution of a stalled replication fork a necessary consequence of replication or should this phenomenon fall under the blanket of DNA repair? Further complicating the matter, many of the same components function in both replication and repair pathways. Nevertheless, it is becoming increasingly apparent that RecQ helicases function in numerous pathways associated with DNA repair and replication (Fig. 4).

Sgs1 has been placed in the same epistasis group as *POL2* (encoding DNA polymerase ε) in a signaling branch that operates parallel to *RAD17/RAD24* in the intra-S phase checkpoint, and Sgs1 localizes to Rad53 foci during S phase.[9] WS cells exhibit prolonged S phase[28] and other abnormalities in DNA replication.[161,162] WRN has been shown to physically interact with a number of components of the replication apparatus including PCNA,[109] Polδ[66] and RPA.[47] Analysis of *srs2 sgs1* conditional mutants indicates a redundant but essential role for these helicases in replication.[136] Sgs1 physically interacts with Rad51. The significance of this interaction has not been established, but it is likely to be physiologically relevant based on the functional interaction between *E. coli* RecQ and RecA.[112] Both WRN and BLM co-localize with Rad51 foci in response to DNA damage.[112,115]

Three separable functions for the eukaryotic RecQ helicases in replication have been proposed: the restart of stalled replication, recombination repair of lesions at stalled forks and the resolution of aberrant DNA secondary structures. Sgs1 co-localizes with Rad53 in S phase specific foci and evidence suggests that Sgs1 is required for proper association of Rad53 with chromatin.[9] This places Sgs1 at the stalled fork. Sgs1 appears to suppress certain types of recombination under normal circumstances and induces recombination in response to DNA damaging conditions.[12,85] Strains lacking *sgs1* exhibit hyper-recombination involving intra-chromosomal[6,10] inter-chromosomal[6,16] illegitimate[7] and homeologous[8] recombination pathways. Despite these hyper-recombination phenotypes, *sgs1* strains fail to initiate an increase in recombination in response to DNA damage,[12] a pathway that is dependent on Rad52.[85] Consistent with this, the recombination defect is not alleviated in diploids relative to haploids. This ploidy effect is generally interpreted as the contribution of homologous recombination.[12] In addition, genetic analysis also reveals an interaction between *SGS1* and *MSH2*. These genes appear to function in independent pathways that suppress homeologous recombination.[8]

Furthermore, in vitro assays have established RecQ family members as promiscuous helicases capable of unwinding a variety of substrates, including synthetic Holliday junctions.[53,54,164] *E. coli* RecQ can initiate recombination in the presence of RecA and single-stranded binding proteins.[165] This allows for the possibility that RecQ performs a dual role in bacterial cells: an initiator of desired recombination events and a suppressor of illegitimate recombination by resolution of irregular DNA structures.

Given the range of known proteins that interact with Sgs1 it is possible to speculate as to the cooperation of these proteins in the resolution of aberrant DNA structures. These include Mgs1, a single strand annealing protein, Rad51, which binds ssDNA and aligns it with duplex DNA, the topoisomerases Top3, Top2 and perhaps Top1 required for decatenation of duplex DNA. Each of these interactions is conserved in humans with WRN showing interactions with WHIP (the Mgs1 homologue), Rad51, and Top1 and BLM showing interaction with Rad51 and Top3α. WRN, BLM and Sgs1 helicases have each been shown to be capable of unwinding G4 DNA. [51,55,56] These DNA structures may form in vivo at regions of ssDNA during DNA replication and their resolution by RecQ helicases may be crucial to the restart of replication.

Telomere Maintenance

There is a growing body of evidence linking eukaryotic RecQ helicases to the maintenance of telomeres.[166] Telomerase adds bases to the ends of chromosomes by reverse transcribing an RNA telomere template.[167] Most somatic cells do not express the catalytic subunit of telomerase and as a result telomeres gradually erode with each round of DNA replication. The short telomere is recognized by the cell as a DNA break.[166,168] The ensuing cascade of events, termed cellular senescence, is associated with cells entering a state of quiescence, the expression of senescence-associated genes, and eventual cell death. It is now widely recognized that telomere shortening is the mitotic clock that limits division capacity of most primary human cell cultures. In recent years it has been proposed that cellular senescence provides a barrier to tumorigenesis and, in humans, likely contributes to the aging of mitotically active tissues and organs.[169]

Immortalization of cells occurs by one of two mechanisms. Most cells bypass senescence by activating telomerase.[170,171] In the other mechanism, observed 5-10% of the time in vivo, a proportion of cells utilize an alternative homologous recombination pathway known as the "ALT" (alternative lengthening of telomeres) pathway.[172] ALT is poorly understood process resulting in extremely long telomeres[173,174] that co-localize with promyelocytic leukemia (PML) bodies.[175,176]

Unlike most mammalian somatic cells, yeast constitutively express telomerase and only senesce when one or more of the yeast telomerase genes are mutated.[177,178] A yeast strain lacking telomerase gradually loses telomeric sequences and senesces after about 70 divisions. As in the human case, rare cells are able to bypass senescence by activating a *RAD52*-dependent telomere-lengthening mechanism.[179,180] There are two such pathways in yeast. In the first, a major recombination pathway generates "type II" telomeres by greatly amplifying the terminal DNA repeats (TG_{1-3} regions) resulting in a structure that is reminiscent of the very long telomeres observed in mammalian ALT cells. A second minor pathway generates "type I" telomeres that are also long but consist of amplified telomere-associated (Y') sequences.[181] Cells that amplify the TG_{1-3} region (Type II) have a growth advantage over cells that amplify the Y' region (Type I).[181]

EST2 encodes the telomerase catalytic subunit in yeast. Deletion of *EST2* in *sgs1* strains gives rise to a two-fold increase in the rate of telomere loss and a ~30% reduction in division capacity compared to *est2 SGS1* strains.[17] Triple *sgs1 est2 rad52* mutants senesce extremely rapidly and no survivors are generated due to the absolute requirement of Rad52 in recombinational telomere lengthening. Expression of WRN in *est2 sgs1* strains can partially rescue the defect in telomere lengthening, suggesting WRN and Sgs1 may have a similar telomeric function.[17] Additional telomeric defects are seen in *sgs1* strains when cells attempt to recover from senescence. Telomerase-deficient yeast cultures lacking *SGS1* persist for longer in senescence and those that do recover always generate telomeres with Y' amplification (Type I).[16-18] The *sgs1* post-senescent survivors grow poorly and tend to arrest in G2/M presumably due to a high frequency of telomere-telomere fusion.[16,17] Deletion of *RAD50* in *sgs1 tlc1* strains has little effect, whereas deletion of *RAD51* prevents cells from recovering from senescence.[16,18] This indicates that Sgs1 acts with Rad50 to generate type II telomeres (TG_{1-3} amplification) and that Rad51 is important for the alternative type I (Y' amplification) default pathway.[16,18,181]

The requirement of Sgs1 for generation of telomeres with TG_{1-3} region amplification (Type II) probably reflects the manner in which these types of structures are generated. The genetic link between Rad50 and Sgs1 in the formation of survivors suggests that these proteins are directly involved in the putative break-induced replication (BIR) mechanism of telomerase-independent telomere lengthening.[182] In BIR, a DNA strand of one chromosome invades a recipient chromosome and the recipient is used as a template for replication to the end of the chromosome.[183] The intermediate structure resembles those that are used to reinitiate replication at stalled forks.[183] Consistent with this model, yeast BIR is *RAD51*-independent,[184] and Sgs1 localizes to stalled replication forks.[9,160]

There is growing evidence that the Sgs1 telomeric function is conserved in humans. Fibroblasts cultured from WS individuals senesce more rapidly in culture than those from normal individuals,[30,31] although this finding has recently been challenged.[32] Ectopic expression of the telomerase catalytic subunit (hTERT) into WS cells can bypass senescence, implying that the accelerated senescence of these cells is due to telomere loss.[185] Biochemical studies also point to a role for RecQ helicases in telomere maintenance. WRN can unwind G4 DNA and other secondary structures in telomeric DNA raising the possibility that WRN may resolve such structures in vivo.[186] Interestingly, WRN interacts directly with Ku70, a protein required for NHEJ and telomere capping.[187-189]

There is also data indicating that WRN participates in recombination-mediated telomere lengthening. First, the rate of spontaneous immortalization of WS cell lines is very low, indicating that they may be defective in recombination- and telomerase-mediated telomere lengthening.[16,185] Second, the ALT pathway is apparently defective in WRN-/- cells.[16] Third, WRN co-localizes with telomeric factor TRF2 and PML nuclear bodies, suggesting that WRN directly participates in ALT.[16,176] Fourth, higher levels of WRN helicase are observed in some transformed cells and tumor cells,[190] possibly to facilitate telomere-telomere recombination. Given that ALT telomeres have been found in immortalized cells and tumors, it is tempting to speculate that the spectrum of tumors in RecQ diseases is shaped by a defect in the ALT pathway.[16-18]

Much of WS pathology occurs in slowly dividing tissues such as the dermis and vascular endothelium and not in bone marrow, epidermis and the gastrointestinal tract.[191,192] This may be explained if telomerase masks the telomere defect in WS cells, as it does in *sgs1* cells. This defect should be most apparent in slowly dividing tissues whose progenitor cells do not express telomerase and least apparent in actively dividing tissues.[16] The reduced replicative capacity of WS cells may also be related to their increased genome instability and apparent defects in DNA replication. Consistent with this is the report that WS cells do not always undergo normal replicative senescence based on their inability to repress c-fos expression during senescence.[193] The reduced proliferative potential of WS cells is most likely the result of two processes: accelerated loss of telomeric sequences and stochastic cell cycle arrest. Both may be different manifestations of the same defect.

Aging

The maintenance of genome stability is crucial for the long-term viability of organisms because mutations and chromosomal rearrangements are usually cumulative. Multiple mechanisms have evolved to combat genome instability including DNA repair/recombination,[194] DNA methylation,[195,196] and the packaging of DNA into a compact nucleosomal structure known as heterochromatin.[197-199] Defects in the maintenance of genome stability lead to a variety of diseases including the RecQ syndromes, tri-nucleotide repeat diseases such as Parkinson's syndrome, Cockayne's syndrome and many types of cancer-susceptibility diseases. Of these, only WS, RTS and to a limited extent Cockayne's syndrome, are characterized by symptoms that resemble premature aging. The implication of these findings is that only certain types of genome instabilities give rise to premature aging phenotypes. This is also true for *S. cerevisiae*.

Aging in *S. cerevisiae* stems primarily from an inability to maintain the stability of the highly repetitive rDNA.[37] Consistent with this, genetic manipulations that stabilize yeast rDNA extend life span significantly,[37,200-203] and those that destabilize rDNA can lead to premature aging.[201,202,204,205] The study of aging in individual yeast cells is made possible by virtue of their asymmetric cell division. A newly formed "daughter" cell can be identified because it is almost always smaller than the "mother" cell that gave rise to it. Mother cells divide about 20 times before dying and exhibit characteristic aging phenotypes. These include an increase in cell size, a slowing of the cell cycle, fragmentation of the nucleolus and sterility due to leaky expression from the heterochromatic mating type locus.[206]

The repeated rDNA locus is localized in the nucleolus and, in *S. cerevisiae*, is comprised of 100-200 tandem copies of a 9 kb rDNA each containing three autonomously replicating sequences (*ARS*s). Stability at this locus is vital for longevity in this organism. During a yeast cell's life span, an apparently stochastic event leads to the excision of an extrachromosomal rDNA circle (ERC) from the rDNA locus via homologous recombination.[37,207] ERCs replicate and tend not to be passed to daughter cells so that after about 15 divisions, the number of ERCs reaches more than 1000 copies. Death ensues, most likely because ERCs titrate essential transcription and/or replication factors.[37] The silent information regulatory complex (Sir2/3/4) forms stabilizing heterochromatin at four loci: the rDNA, telomeres and the two transcriptionally "silent" mating type loci.[208] During aging, Sir proteins relocalize from telomeres and *HM* loci to the nucleolus,[209] perhaps in an attempt to stem the tide of ERCs.[37] The result is an inability to maintain silenced regions at the mating type locus and mating-type genes are expressed, causing the cell to enter a pseudo-diploid, sterile state.[210]

A loss-of-function mutation in *SGS1* leads to a 60% reduction in life span.[11,14,20,129,135] Old *sgs1* cells exhibit the hallmarks of premature aging, including an increase in cell size, a slowing of the cell cycle, the movement of Sir3 to the nucleolus and sterility.[20] Not all yeast DNA repair genes reduce life span when mutated. For example, mutations in NER, SSA or NHEJ have no effect.[201,204] Sgs1 helicase activity is essential for a normal life span,[13,14,135] but not for resistance to topoisomerase inhibitors.[13] Interestingly, the C-terminus of Sgs1, which facilitates the Rad51 interaction, is fully dispensable for a normal life span.[211] BLM but not WRN can complement the short life span of *sgs1* strains.[100] The complementation of life span is reflected in the ability of BLM to rescue the hyper-recombination of *sgs1* strains at the rDNA locus.

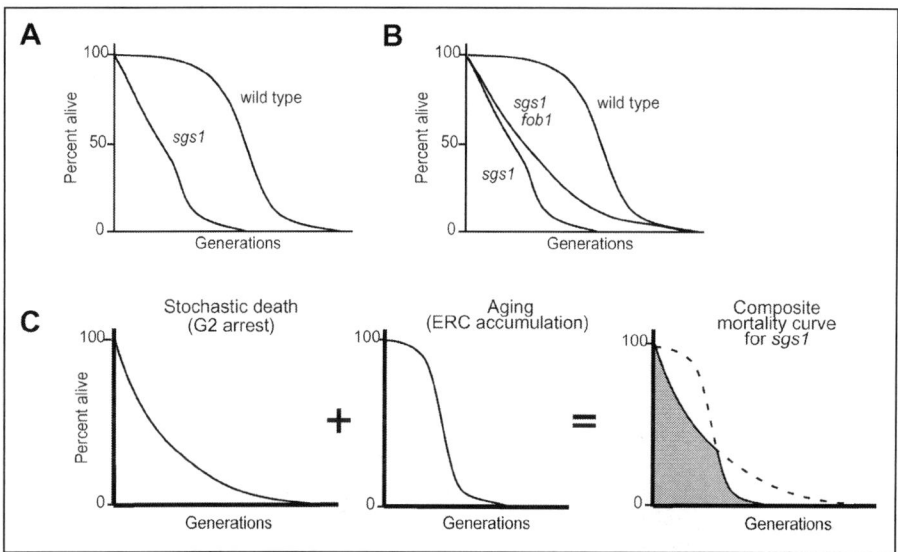

Figure 5. The short life span of sgs1 cells is a composite of aging and stochastic death. Yeast exhibit characteristic aging phenotypes such as increased cell size, a slowing of the cell cycle, fragmentation of the nucleolus, the relocalization of the Sir complex (silent-information regulator) from telomeres and HM loci to the nucleolus, resulting in sterility. One of the causes of yeast aging involves the accumulation of extrachromosomal ribosomal DNA circles (ERCs) to toxic levels, killing the cell. A) For a given population of wild type cells, a plot of the percent of cells alive at each age produces a characteristic Gompertz curve, with a shoulder at young ages due to the minimal impact of ERCs in young cells. Mutations that destabilize the rDNA locus (e.g., sir2D) usually cause premature aging whereas those that stabilize the rDNA delay aging and extend life span. Deletion of FOB1 in wild type strains reduces ERC formation and extends life span by up to 50%. Loss-of-function mutations in SGS1 cause a 60% reduction in life span and old sgs1 cells exhibit the hallmarks of premature aging. B) The maximum life span of sgs1 cells is extended by a fob1 deletion. Thus, cells that die at later generations are subject to aging at a premature stage. The lack of effect of fob1 earlier in the curve indicates that these cells die as a result of non-aging related processes. C) The sgs1 life span curve is a composite of a stochastic process leading to G2 arrest predominately at young ages and the effect of aging due to ERC accumulation predominately at later ages.

Recent studies of *sgs1* cells indicate that the short life span of *sgs1* cells is composite of two independent processes: aging and stochastic death (Fig. 5). Evidence for the aging component comes from epistasis analysis between *SGS1* and *FOB1*. Deleting *FOB1*, which encodes an rDNA replication fork-block protein, stabilizes the rDNA and delays the formation of ERCs.[129,200] Deletion of *FOB1* significantly increases the maximum life span of *sgs1* strains. The non-aging, stochastic component of the *sgs1* life span appears to stem from defects in DNA replication, leading to cell cycle arrest in G2/M. Consistent with the two component model, deletion of *FOB1* has little effect on the death rate of young *sgs1* cells because this is when stochastic death exerts its greatest impact—prior to the effect of accumulating ERCs.[129]

Strains lacking *SRS2* have also been shown to have a short life span that is the outcome of aging and stochastic death.[129] Interestingly, overexpression of *SGS1* in *srs2* strains suppresses hydroxyurea and MMS sensitivity and increases the maximum but not the average life span.[135] This effect is almost identical to that of deleting *FOB1* in an *sgs1* strain, which has the effect of delaying aging (see Fig. 5). Thus, it is tempting to speculate that the increase in maximum life

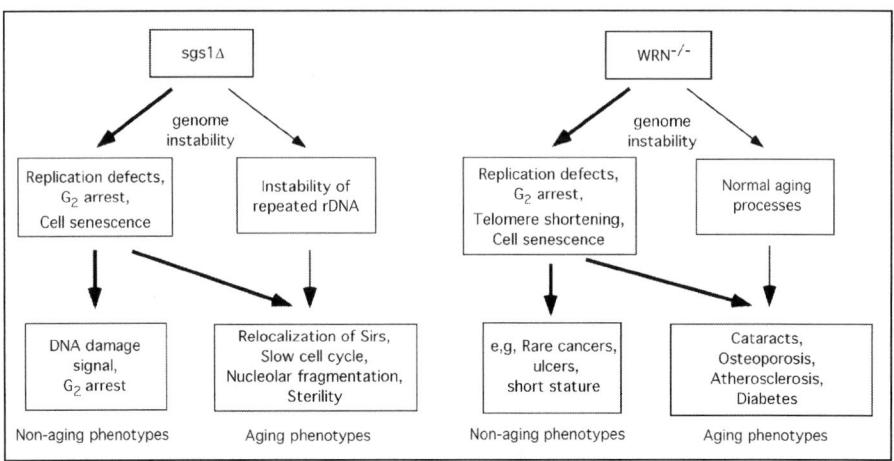

Figure 6. Model to explain sgs1 and WS phenotypes. Defects in Sgs1 or WRN lead to genome instability that is manifested as premature aging and non-aging phenotypes. In *S. cerevisiae*, a major cause of aging in normal cells stems from the instability of the repeated ribosomal DNA (rDNA) locus. Aging in yeast is accompanied by a slowing of the cell cycle, fragmentation of the nucleolus, the relocalization of the Sir complex from telomeres and mating-type loci to the nucleolus, resulting in a sterile, pseudo-diploid state. Life span in sgs1 mutants is a composite of normal aging and stochastic death due to defects in DNA replication (see Fig. 4). Similarly, WS is the consequence of aging- and non-aging-related symptoms stemming from different aspects of genome instability. Non-aging related processes in both species can give rise to phenotypes that resemble normal aging.

span in *srs2* strains is also due to the suppression of aging but not the apparent DNA replication defect.

ERCs and other circular DNAs are readily found in most mammalian cell types, although they do not appear to correlate with the age of the donor or passage number. However, mammals possess many more regions of repetitive DNA than yeast, and if ERCs were to occur from all of these loci, not just the rDNA, they would be difficult to detect.[3] Nevertheless, it is important to note that aging is not an adaptive biological process (like DNA repair, for example) and is therefore not necessarily conserved.[212] The fact that ERCs limit life span in yeast indicates that the maintenance of genome stability is vital for the longevity of the organism. This is probably true for all eukaryotes. Genome instability likely manifests itself in different ways according to the biology of the organism. Analysis of the shortened life span of *sgs1* mutants reveals non-aging and aging components, both of which stem from genome instability. WS phenotypes can be understood in the same terms, as indicated in Figure 6.

Meiosis

Sgs1 has a defect in the first division of meiosis[6] and meiotic recombination is reduced in *sgs1* strains.[19] Interestingly the helicase activity is not required for meiotic function in yeast. This ties in with the fact that this activity is not required for interaction with Top1 or Top3 [10,213] or for resistance to topoisomerase inhibitors.[14] Top3 and Sgs1 form a stable complex through the N-terminal region of Sgs1,[10,102,163] and the interaction is independent of DNA. It is possible that Sgs1 recruits Top3 to the DNA under normal circumstances but that Top3 has an innately inefficient mechanism to load itself on to DNA. This Sgs1 function appears to be conserved for BLM but not WRN. Male BS patients are sterile[33] and studies in mouse indicate that BLM, Top3α and Top3β expression is upregulated in meiosis.[214] BLM associates with meiotic chromosomes during meiosis where it co-localizes with RPA.[34] No role for WRN in meiosis has been reported to date and sterility does not appear to be a common symptom of WS.[215]

Summary

An enormous wealth of information about the yeast RecQ has been obtained through genetic analysis. This has lead to the identification of interacting genes and the elucidation of pathways involved in DNA replication, repair and recombination. This type of analysis has also revealed the relative contribution from the various genes and pathways that are vital to the genomic stability of the organism. In terms of the biochemical aspects of RecQ helicase function, yeast models have been employed to a lesser extent, and in this context more has been done in human cell culture. It is likely that new and relevant data could be gained from further biochemical analysis and purification of Sgs1 complexes.

Although a substantial amount of data on the RecQ helicases comes from in vitro analyses, there are obvious limitations to the type of conclusions that may be drawn from in vitro work. Under physiological conditions RecQ proteins are almost certainly never unaccompanied. As the biochemical studies have shown, interactions with other proteins likely alter substrate specificity and activity of the helicase. Moreover, it is clear that the post-transcriptional modification of RecQ proteins by phosphorylation and sumoylation can modulate the catalytic activity and localization of these proteins. A complete picture of RecQ function will only emerge from a combination of genetic, cell biological and biochemical studies. Much of the speculation on the function of the eukaryotic RecQ proteins has relied on clues from prokaryotes. Replication, repair and recombination have been studied extensively in *E. coli* and much of what we know today stems from pioneering work in this organism. However, there are a large number of fundamental differences between bacteria and eukaryotes that limit the usefulness of such models. Yeast has certainly proven itself to be a useful model of RecQ diseases, especially in identifying genetic interactions. A comparison of Sgs1 functions and mutant phenotypes with those of WRN illustrates how closely related they are. The fact that Sgs1 is the sole RecQ helicase in budding yeast indicates that it likely performs multiple functions that have been divided between the human RecQ helicases. Indeed, some aspects of Sgs1 function are more applicable to BLM than to WRN. Nevertheless, the ease with which yeast can be manipulated genetically means that yeast will likely provide additional clues to the functions of WRN and the other human RecQ helicases for many years to come.

References

1. Kowalczykowski SC, Dixon DA, Eggleston AK et al. Biochemistry of homologous recombination in Escherichia coli. Microbiol Rev 1994; 58:401-465.
2. Courcelle J, Hanawalt PC. Participation of recombination proteins in rescue of arrested replication forks in UV-irradiated Escherichia coli need not involve recombination. Proc Natl Acad Sci USA 2001; 98:8196-8202.
3. Shen J, Loeb LA. Unwinding the molecular basis of the Werner syndrome. Mech Ageing Dev 2001; 122:921-944.
4. Gangloff S, McDonald JP, Bendixen C et al. The yeast type I topoisomerase Top3 interacts with Sgs1, a DNA helicase homolog: a potential eukaryotic reverse gyrase. Mol Cell Biol 1994; 14:8391-8398.
5. Watt PM, Louis EJ, Borts RH et al. Sgs1: A eukaryotic homolog of E. coli RecQ that interacts with topoisomerase II in vivo and is required for faithful chromosome segregation. Cell 1995; 81:253-260.
6. Watt PM, Hickson ID, Borts RH et al. SGS1, a homologue of the Bloom's and Werner's syndrome genes, is required for maintenance of genome stability in Saccharomyces cerevisiae. Genetics 1996; 144:935-945.

7. Yamagata K, Kato J, Shimamoto A et al. Bloom's and Werner's syndrome genes suppress hyperrecombination in yeast sgs1 mutant: implication for genomic instability in human diseases. Proc Natl Acad Sci USA 1998; 95:8733-8738.
8. Myung K, Datta A, Chen C et al. SGS1, the Saccharomyces cerevisiae homologue of BLM and WRN, suppresses genome instability and homeologous recombination. Nat Genet 2001; 27:113-116.
9. Frei C, Gasser SM. The yeast Sgs1p helicase acts upstream of Rad53p in the DNA replication checkpoint and colocalizes with Rad53p in S-phase-specific foci. Genes Dev 2000; 14:81-96.
10 Mullen JR, Kaliraman V, Brill SJ. Bipartite structure of the SGS1 DNA helicase in Saccharomyces cerevisiae. Genetics 2000; 154:1101-1114.
11. Saffi J, Feldmann H, Winnacker EL et al. Interaction of the yeast Pso5/Rad16 and Sgs1 proteins: influences on DNA repair and aging. Mutat Res 2001; 486:195-206.
12. Gangloff S, Soustelle C, Fabre F, et al. Homologous recombination is responsible for cell death in the absence of the Sgs1 and Srs2 helicases. Nat Genet 2000; 25:192-194.
13. Saffi J, Pereira VR, Henriques JA. Importance of the Sgs1 helicase activity in DNA repair of Saccharomyces cerevisiae. Curr Genet 2000; 37:75-78.
14. Mankouri HW, Morgan A. The DNA helicase activity of yeast Sgs1p is essential for normal lifespan but not for resistance to topoisomerase inhibitors. Mech Ageing Dev 2001; 122:1107-1120.
15. Hishida T, Iwasaki H, Ohno T et al. A yeast gene, MGS1, encoding a DNA-dependent AAA(+) ATPase is required to maintain genome stability. Proc Natl Acad Sci USA 2001; 98:8283-8289.
16. Johnson FB, Marciniak RA, McVey M et al. The Saccharomyces cerevisiae WRN homolog Sgs1p participates in telomere maintenance in cells lacking telomerase. Embo J 2001; 20:905-913.
17. Cohen H, Sinclair DA. Recombination-mediated lengthening of terminal telomeric repeats requires the Sgs1 DNA helicase. Proc Natl Acad Sci USA 2001; 98:3174-3179.
18. Huang P, Pryde FE, Lester D et al. SGS1 is required for telomere elongation in the absence of telomerase. Curr Biol 2001; 11:125-129.
19. Miyajima A, Seki M, Onoda F et al. Sgs1 helicase activity is required for mitotic but apparently not for meiotic functions. Mol Cell Biol 2000; 20:6399-6409.
20. Sinclair DA, Mills K, Guarente L. Accelerated aging and nucleolar fragmentation in yeast sgs1 mutants. Science 1997; 277:1313-1316.
21. Krepinsky AB, Rainbow AJ, Heddle JA. Studies on the ultraviolet light sensitivity of Bloom's syndrome fibroblasts. Mutat Res 1980; 69:357-368.
22. Evans HJ, Adams AC, Clarkson JM et al. Chromosome aberrations and unscheduled DNA synthesis in X- and UV- irradiated lymphocytes from a boy with Bloom's syndrome and a man with xeroderma pigmentosum. Cytogenet Cell Genet 1978; 20:124-140.
23. Salk D, Au K, Hoehn H et al. Evidence of clonal attenuation, clonal succession, and clonal expansion in mass cultures of aging Werner's syndrome skin fibroblasts. Cytogenet Cell Genet 1981; 30:108-117.
24. Fukuchi K, Martin GM, Monnat Jr RJ. Mutator phenotype of Werner syndrome is characterized by extensive deletions. Proc Natl Acad Sci USA 1989; 86:5893-5897.
25. Lebel M. Increased frequency of DNA deletions in pink-eyed unstable mice carrying a mutation in the Werner syndrome gene homologue. Carcinogenesis 2002; 23:213-216.
26. Pichierri P, Franchitto A, Piergentili R et al. Hypersensitivity to camptothecin in MSH2 deficient cells is correlated with a role for MSH2 protein in recombinational repair. Carcinogenesis 2001; 22:1781-1787.
27. Lebel M, Leder P. A deletion within the murine Werner syndrome helicase induces sensitivity to inhibitors of topoisomerase and loss of cellular proliferative capacity. Proc Natl Acad Sci USA 1998; 95:13097-13102.
28. Poot M, Gollahon KA, Rabinovitch PS. Werner syndrome lymphoblastoid cells are sensitive to camptothecin- induced apoptosis in S-phase. Hum Genet 1999; 104:10-14.
29. Gebhart E, Bauer R, Raub U et al. Spontaneous and induced chromosomal instability in Werner syndrome. Hum Genet 1988; 80:135-139.
30. Tahara H, Tokutake Y, Maeda S et al. Abnormal telomere dynamics of B-lymphoblastoid cell strains from Werner's syndrome patients transformed by Epstein-Barr virus. Oncogene 1997; 15:1911-1920.

31. Schulz VP, Zakian VA, Ogburn CE et al. Accelerated loss of telomeric repeats may not explain accelerated replicative decline of Werner syndrome cells. Hum Genet 1996; 97:750-754.
32. Choi D, Whittier PS, Oshima J et al. Telomerase expression prevents replicative senescence but does not fully reset mRNA expression patterns in Werner syndrome cell strains. Faseb J 2001; 15:1014-1020.
33. German J. Bloom syndrome: A mendelian prototype of somatic mutational disease. Medicine (Baltimore) 1993; 72:393-406.
34. Walpita D, Plug AW, Neff NF et al. Bloom's syndrome protein, BLM, colocalizes with replication protein A in meiotic prophase nuclei of mammalian spermatocytes. Proc Natl Acad Sci USA 1999; 96:5622-5627.
35. Epstein CJ, Martin GM, Schultz AL et al. Werner's syndrome a review of its symptomatology, natural history, pathologic features, genetics and relationship to the natural aging process. Medicine (Baltimore) 1966; 45:177-221.
36. Martin GM. Genetics and aging; the Werner syndrome as a segmental progeroid syndrome. Adv Exp Med Biol 1985; 190:161-170.
37. Sinclair DA, Guarente L. Extrachromosomal rDNA circles—a cause of aging in yeast. Cell 1997; 91:1033-1042.
38. Stewart E, Chapman CR, Al-Khodairy F et al. rqh1+, a fission yeast gene related to the Bloom's and Werner's syndrome genes, is required for reversible S phase arrest. Embo J 1997; 16:2682-2692.
39. Davey S, Han CS, Ramer SA et al. Fission yeast rad12+ regulates cell cycle checkpoint control and is homologous to the Bloom's syndrome disease gene. Mol Cell Biol 1998; 18:2721-2728.
40. Murray JM, Lindsay HD, Munday CA et al. Role of Schizosaccharomyces pombe RecQ homolog, recombination, and checkpoint genes in UV damage tolerance. Mol Cell Biol 1997; 17:6868-6875.
41. Hofmann AF, Harris SD. The Aspergillus nidulans musN gene encodes a RecQ helicase that interacts with the PI-3K-related kinase UVSB. Genetics 2001; 159:1595-1604.
42. Cogoni C, Macino G. Posttranscriptional gene silencing in Neurospora by a RecQ DNA helicase. Science 1999; 286:2342-2344.
43. Irino N, Nakayama K, Nakayama H. The recQ gene of Escherichia coli K12: Primary structure and evidence for SOS regulation. Mol Gen Genet 1986; 205:298-304.
44. Karow JK, Wu L, Hickson ID. RecQ family helicases: roles in cancer and aging. Curr Opin Genet Dev 2000; 10:32-38.
45. Huang S, Beresten S, Li B et al. Characterization of the human and mouse WRN 3'—>5' exonuclease. Nucleic Acids Res 2000; 28:2396-2405.
46. Huang S, Li B, Gray MD et al. The premature ageing syndrome protein, WRN, is a 3'—>5' exonuclease. Nat Genet 1998; 20:114-116.
47. Shen JC, Gray MD, Oshima J et al. Characterization of Werner syndrome protein DNA helicase activity: directionality, substrate dependence and stimulation by replication protein A. Nucleic Acids Res 1998; 26:2879-2885.
48. Suzuki N, Shiratori M, Goto M et al. Werner syndrome helicase contains a 5'—>3' exonuclease activity that digests DNA and RNA strands in DNA/DNA and RNA/DNA duplexes dependent on unwinding. Nucleic Acids Res 1999; 27:2361-2368.
49. Bennett RJ, Sharp JA, Wang JC. Purification and characterization of the Sgs1 DNA helicase activity of Saccharomyces cerevisiae. J Biol Chem 1998; 273:9644-9650.
50. Bennett RJ, Keck JL, Wang JC. Binding specificity determines polarity of DNA unwinding by the Sgs1 protein of S. cerevisiae. J Mol Biol 1999; 289:235-248.
51. Sun H, Bennett RJ, Maizels N. The Saccharomyces cerevisiae Sgs1 helicase efficiently unwinds G-G paired DNAs. Nucleic Acids Res 1999; 27:1978-1984.
52. Han H, Bennett RJ, Hurley LH. Inhibition of unwinding of G-quadruplex structures by sgs1 helicase in the presence of N,N'-Bis[2-(1-piperidino)ethyl]-3,4,9, 10- perylenetetracarboxylic diimide, a G-quadruplex-interactive ligand. Biochemistry 2000; 39:9311-9316.
53. Constantinou A, Tarsounas M, Karow JK et al. Werner's syndrome protein (WRN) migrates Holliday junctions and co-loclaizes with RPA upon replication arrest. EMBO Reports 2000; 1:80-84.
54. Karow JK, Constantinou A, Li JL et al. The Bloom's syndrome gene product promotes branch migration of holliday junctions. Proc Natl Acad Sci USA 2000; 97:6504-6508.

55. Fry M, Loeb LA. Human werner syndrome DNA helicase unwinds tetrahelical structures of the fragile X syndrome repeat sequence d(CGG)n. J Biol Chem 1999; 274:12797-12802.
56. Sun H, Karow JK, Hickson ID et al. The Bloom's syndrome helicase unwinds G4 DNA. J Biol Chem 1998; 273:27587-27592.
57. Venczel EA, Sen D. Parallel and antiparallel G-DNA structures from a complex telomeric sequence. Biochemistry 1993; 32:6220-6228.
58. Muniyappa K, Anuradha S, Byers B. Yeast meiosis-specific protein Hop1 binds to G4 DNA and promotes its formation. Mol Cell Biol 2000; 20:1361-1369.
59. Liu Z, Lee A, Gilbert W. Gene disruption of a G4-DNA-dependent nuclease in yeast leads to cellular senescence and telomere shortening. Proc Natl Acad Sci USA 1995; 92:6002-6006.
60. Tishkoff DX, Rockmill B, Roeder GS et al. The sep1 mutant of Saccharomyces cerevisiae arrests in pachytene and is deficient in meiotic recombination. Genetics 1995; 139:495-509.
61. Liu Z, Gilbert W. The yeast KEM1 gene encodes a nuclease specific for G4 tetraplex DNA: implication of in vivo functions for this novel DNA structure. Cell 1994; 77:1083-1092.
62. Patel SS, Picha KM. Structure and function of hexameric helicases. Annu Rev Biochem 2000; 69:651-697.
63. Fricke WM, Kaliraman V, Brill SJ. Mapping the DNA topoisomerase III binding domain of the Sgs1 DNA helicase. J Biol Chem 2001; 276:8848-8855.
64. Gray MD, Wang L, Youssoufian H et al. Werner helicase is localized to transcriptionally active nucleoli of cycling cells. Exp Cell Res 1998; 242:487-494.
65. Marciniak RA, Lombard DB, Johnson FB et al. Nucleolar localization of the Werner syndrome protein in human cells. Proc Natl Acad Sci USA 1998; 95:6887-6892.
66. Szekely AM, Chen YH, Zhang C et al. Werner protein recruits DNA polymerase delta to the nucleolus. Proc Natl Acad Sci USA 2000; 97:11365-11370.
67. Yannone SM, Roy S, Chan DW et al. Werner syndrome protein is regulated and phosphorylated by DNA- dependent protein kinase. J Biol Chem 2001; 276:38242-38248.
68. Karmakar P, Piotrowski J, Brosh Jr RM et al. Werner protein is a target of DNA-PK in vivo and in vitro and its catalytic activities are regulated by phosphorylation. J Biol Chem 2002; 11:11
69. Dutertre S, Ababou M, Onclercq R et al. Cell cycle regulation of the endogenous wild type Bloom's syndrome DNA helicase. Oncogene 2000; 19:2731-2738.
70. Franchitto A, Pichierri P. Bloom's syndrome protein is required for correct relocalization of RAD50/MRE11/NBS1 complex after replication fork arrest. J Cell Biol 2002; 157:19-30.
71. Ababou M, Dutertre S, Lecluse Y et al. ATM-dependent phosphorylation and accumulation of endogenous BLM protein in response to ionizing radiation. Oncogene 2000; 19:5955-5963.
72. Shimamoto A, Nishikawa K, Kitao S et al. Human RecQ5beta, a large isomer of RecQ5 DNA helicase, localizes in the nucleoplasm and interacts with topoisomerases 3alpha and 3beta. Nucleic Acids Res 2000; 28:1647-1655.
73. Goodwin A, Wang SW, Toda T et al. Topoisomerase III is essential for accurate nuclear division in Schizosaccharomyces pombe. Nucleic Acids Res 1999; 27:4050-4058.
74. Pedrazzi G, Perrera C, Blaser H et al. Direct association of Bloom's syndrome gene product with the human mismatch repair protein MLH1. Nucleic Acids Res 2001; 29:4378-4386.
75. Moser MJ, Oshima J, Monnat Jr RJ. WRN mutations in Werner syndrome. Hum Mutat 1999; 13:271-279.
76. Monnat Jr RJ. Werner syndrome: molecular genetics and mechanistic hypotheses. Exp Gerontol 1992; 27:447-453.
77. Salk D, Au K, Hoehn H et al. Cytogenetic aspects of Werner syndrome. Adv Exp Med Biol 1985; 190:541-546.
78. Monnat Jr RJ, Hackmann AF, Chiaverotti TA. Nucleotide sequence analysis of human hypoxanthine phosphoribosyltransferase (HPRT) gene deletions. Genomics 1992; 13:777-787.
79. Hanada K, Ukita T, Kohno Y et al. RecQ DNA helicase is a suppressor of illegitimate recombination in Escherichia coli. Proc Natl Acad Sci USA 1997; 94:3860-3865.
80. Wang Y, Cortez D, Yazdi P et al. BASC, a super complex of BRCA1-associated proteins involved in the recognition and repair of aberrant DNA structures. Genes Dev 2000; 14:927-939.
81. Alani E, Lee S, Kane MF et al. Saccharomyces cerevisiae MSH2, a mispaired base recognition protein, also recognizes Holliday junctions in DNA. J Mol Biol 1997; 265:289-301.

82. Marsischky GT, Lee S, Griffith J et al. 'Saccharomyces cerevisiae MSH2/6 complex interacts with Holliday junctions and facilitates their cleavage by phage resolution enzymes. J Biol Chem 1999; 274:7200-7206.
83. Kawabe Y, Branzei D, Hayashi T et al. A novel protein interacts with the Werner's syndrome gene product physically and functionally. J Biol Chem 2001; 276:20364-20369.
84. Matsumoto Y. Molecular mechanism of PCNA-dependent base excision repair. Prog Nucleic Acid Res Mol Biol 2001; 68:129-138.
85. Onoda F, Seki M, Miyajima A et al. Involvement of SGS1 in DNA damage-induced heteroallelic recombination that requires RAD52 in Saccharomyces cerevisiae. Mol Gen Genet 2001; 264:702-708.
86. Champoux JJ. DNA topoisomerases: Structure, function, and mechanism. Annu Rev Biochem 2001; 70:369-413.
87. Kim RA, Wang JC. A subthreshold level of DNA topoisomerases leads to the excision of yeast rDNA as extrachromosomal rings. Cell 1989; 57:975-985.
88. Castano IB, Brzoska PM, Sadoff BU et al. Mitotic chromosome condensation in the rDNA requires TRF4 and DNA topoisomerase I in Saccharomyces cerevisiae. Genes Dev 1996; 10:2564-2576.
89. Choder M. A general topoisomerase I-dependent transcriptional repression in the stationary phase in yeast. Genes Dev 1991; 5:2315-2326.
90. Goto T, Wang JC. Cloning of yeast TOP1, the gene encoding DNA topoisomerase I, and construction of mutants defective in both DNA topoisomerase I and DNA topoisomerase II. Proc Natl Acad Sci USA 1985; 82:7178-7182.
91. Thrash C, Bankier AT, Barrell BG et al. Cloning, characterization, and sequence of the yeast DNA topoisomerase I gene. Proc Natl Acad Sci USA 1985; 82:4374-4378.
92. Holm C, Goto T, Wang JC et al. DNA topoisomerase II is required at the time of mitosis in yeast. Cell 1985; 41:553-563.
93. Holm C, Stearns T, Botstein D. DNA topoisomerase II must act at mitosis to prevent nondisjunction and chromosome breakage. Mol Cell Biol 1989; 9:159-168.
94. Uemura T, Ohkura H, Adachi Y et al. DNA topoisomerase II is required for condensation and separation of mitotic chromosomes in S. pombe. Cell 1987; 50:917-925.
95. Kim RA, Wang JC. Identification of the yeast TOP3 gene product as a single strand- specific DNA topoisomerase. J Biol Chem 1992; 267:17178-17185.
96. Wallis JW, Chrebet G, Brodsky G et al. A hyper-recombination mutation in S. cerevisiae identifies a novel eukaryotic topoisomerase. Cell 1989; 58:409-419.
97. Chakraverty RK, Kearsey JM, Oakley TJ et al. Topoisomerase III acts upstream of Rad53p in the S-phase DNA damage checkpoint. Mol Cell Biol 2001; 21:7150-7162.
98. Bailis AM, Arthur L, Rothstein R. Genome rearrangement in top3 mutants of Saccharomyces cerevisiae requires a functional RAD1 excision repair gene. Mol Cell Biol 1992; 12:4988-4993.
99. Gangloff S, de Massy B, Arthur L et al. The essential role of yeast topoisomerase III in meiosis depends on recombination. Embo J 1999; 18:1701-1711.
100. Heo SJ, Tatebayashi K, Ohsugi I et al. Bloom's syndrome gene suppresses premature ageing caused by Sgs1 deficiency in yeast. Genes Cells 1999; 4:619-625.
101. Duno M, Thomsen B, Westergaard O et al. Genetic analysis of the Saccharomyces cerevisiae Sgs1 helicase defines an essential function for the Sgs1-Top3 complex in the absence of SRS2 or TOP1. Mol Gen Genet 2000; 264:89-97.
102. Bennett RJ, Noirot-Gros MF, Wang JC. Interaction between yeast Sgs1 helicase and DNA topoisomerase III. J Biol Chem 2000; 275:26898-26905.
103. Bennett RJ, Wang JC. Association of yeast DNA topoisomerase III and Sgs1 DNA helicase: studies of fusion proteins. Proc Natl Acad Sci USA 2001; 98:11108-11113.
104. Hu P, Beresten SF, van Brabant AJ et al. Evidence for BLM and Topoisomerase IIIalpha interaction in genomic stability. Hum Mol Genet 2001; 10:1287-1298.
105. Harmon FG, DiGate RJ, Kowalczykowski SC. RecQ helicase and topoisomerase III comprise a novel DNA strand passage function: A conserved mechanism for control of DNA recombination. Mol Cell 1999; 3:611-620.
106. Wu L, Davies SL, North PS et al. The Bloom's syndrome gene product interacts with topoisomerase III. J Biol Chem 2000; 275 9636-9644.

107. Ng SW, Liu Y, Hasselblatt KT et al. A new human topoisomerase III that interacts with SGS1 protein. Nucleic Acids Res 1999; 27:993-1000.
108. Confalonieri F, Elie C, Nadal M et al. Reverse gyrase: A helicase-like domain and a type I topoisomerase in the same polypeptide. Proc Natl Acad Sci USA 1993; 90:4753-4757.
109. Lebel M, Spillare EA, Harris CC et al. The Werner syndrome gene product co-purifies with the DNA replication complex and interacts with PCNA and topoisomerase I. J Biol Chem 1999; 274:37795-37799.
110. Baumann P, West SC. Role of the human RAD51 protein in homologous recombination and double-stranded-break repair. Trends Biochem Sci 1998; 23:247-251.
111. Bianco PR, Tracy RB, Kowalczykowski SC. DNA strand exchange proteins: a biochemical and physical comparison. Front Biosci 1998; 3:D570-603.
112. Wu L, Davies SL, Levitt NC et al. Potential role for the BLM helicase in recombinational repair via a conserved interaction with RAD51. J Biol Chem 2001; 276:19375-19381.
113. Bischof O, Kim SH, Irving J et al. Regulation and localization of the Bloom syndrome protein in response to DNA damage. J Cell Biol 2001; 153:367-380.
114. Moens PB, Freire R, Tarsounas M et al. Expression and nuclear localization of BLM, a chromosome stability protein mutated in Bloom's syndrome, suggest a role in recombination during meiotic prophase. J Cell Sci 2000; 113:663-672.
115. Sakamoto S, Nishikawa K, Heo SJ et al. Werner helicase relocates into nuclear foci in response to DNA damaging agents and co-localizes with RPA and Rad51. Genes Cells 2001; 6:421-430.
116. Pichierri P, Franchitto A, Mosesso P et al. Werner's syndrome protein is required for correct recovery after replication arrest and DNA damage induced in S-phase of cell cycle. Mol Biol Cell 2001; 12:2412-2421.
117. Prakash S, Prakash L. Nucleotide excision repair in yeast. Mutat Res 2000; 451:13-24.
118. Guzder SN, Sung P, Prakash L et al. The DNA-dependent ATPase activity of yeast nucleotide excision repair factor 4 and its role in DNA damage recognition. J Biol Chem 1998; 273:6292-6296.
119. Guzder SN, Sung P, Prakash L et al. Yeast Rad7-Rad16 complex, specific for the nucleotide excision repair of the nontranscribed DNA strand, is an ATP-dependent DNA damage sensor. J Biol Chem 1997; 272:21665-21668.
120. Mullen JR, Kaliraman V, Ibrahim SS et al. Requirement for three novel protein complexes in the absence of the Sgs1 DNA helicase in Saccharomyces cerevisiae. Genetics 2001; 157:103-118.
121. Kaliraman V, Mullen JR, Fricke WM et al. Functional overlap between Sgs1-Top3 and the Mms4-Mus81 endonuclease. Genes Dev 2001; 15:2730-2740.
122. Boddy MN, Lopez-Girona A, Shanahan P et al. Damage tolerance protein Mus81 associates with the FHA1 domain of checkpoint kinase Cds1. Mol Cell Biol 2000; 20:8758-8766.
123. Hollingsworth N. unpublished data.
124. Bardwell AJ, Bardwell L, Tomkinson AE et al. Specific cleavage of model recombination and repair intermediates by the yeast Rad1-Rad10 DNA endonuclease. Science 1994; 265:2082-2085.
125. Rong L, Klein HL. Purification and characterization of the SRS2 DNA helicase of the yeast Saccharomyces cerevisiae. J Biol Chem 1993; 268:1252-1259.
126. Fabre F, Magana-Schwencke N, Chanet R. Isolation of the RAD18 gene of Saccharomyces cerevisiae and construction of rad18 deletion mutants. Mol Gen Genet 1989; 215:425-430.
127. Rong L, Palladino F, Aguilera A et al. The hyper-gene conversion hpr5-1 mutation of Saccharomyces cerevisiae is an allele of the SRS2/RADH gene. Genetics 1991; 127:75-85.
128. Schiestl RH, Prakash S, Prakash L. The SRS2 suppressor of rad6 mutations of Saccharomyces cerevisiae acts by channeling DNA lesions into the RAD52 DNA repair pathway. Genetics 1990; 124:817-831.
129. McVey M, Kaeberlein M, Tissenbaum HA et al. The short life span of Saccharomyces cerevisiae sgs1 and srs2 mutants is a composite of normal aging processes and mitotic arrest due to defective recombination. Genetics 2001; 157:1531-1542.
130. Aboussekhra A, Chanet R, Adjiri A et al. Semidominant suppressors of Srs2 helicase mutations of Saccharomyces cerevisiae map in the RAD51 gene, whose sequence predicts a protein with similarities to procaryotic RecA proteins. Mol Cell Biol 1992; 12:3224-3234.
131. Milne GT, Ho T, Weaver DT. Modulation of Saccharomyces cerevisiae DNA double-strand break repair by SRS2 and RAD51. Genetics 1995; 139:1189-1199.

132. Schild D. Suppression of a new allele of the yeast RAD52 gene by overexpression of RAD51, mutations in srs2 and ccr4, or mating-type heterozygosity. Genetics 1995; 140:115-127.
133. Chanet R, Heude M, Adjiri A et al. Semidominant mutations in the yeast Rad51 protein and their relationships with the Srs2 helicase. Mol Cell Biol 1996; 16:4782-4789.
134. Liberi G, Chiolo I, Pellicioli A et al. Srs2 DNA helicase is involved in checkpoint response and its regulation requires a functional Mec1-dependent pathway and Cdk1 activity. Embo J 2000; 19:5027-5038.
135. Mankouri HW, Craig TJ, Morgan A. SGS1 is a multicopy suppressor of srs2: functional overlap between DNA helicases. Nucleic Acids Res 2002; 30:1103-1113.
136. Lee SK, Johnson RE, Yu SL et al. Requirement of yeast SGS1 and SRS2 genes for replication and transcription. Science 1999; 286:2339-2342.
137. Paull TT. New glimpses of an old machine. Cell 2001; 107:563-565.
138. Petrini JH. The mammalian Mre11-Rad50-nbs1 protein complex: Integration of functions in the cellular DNA-damage response. Am J Hum Genet 1999; 64:1264-1269.
139. Sugawara N, Paques F, Colaiacovo M et al. Role of Saccharomyces cerevisiae Msh2 and Msh3 repair proteins in double-strand break-induced recombination. Proc Natl Acad Sci USA 1997; 94:9214-9219.
140. Wassmann K, Benezra R. Mitotic checkpoints: From yeast to cancer. Curr Opin Genet Dev 2001; 11:83-90.
141. Lowndes NF, Murguia JR. Sensing and responding to DNA damage. Curr Opin Genet Dev 2000; 10:17-25.
142. Hartwell L, Weinert T, Kadyk L et al. Cell cycle checkpoints, genomic integrity, and cancer. Cold Spring Harb Symp Quant Biol 1994; 59:259-263.
143. Larner JM, Lee H, Hamlin JL. Radiation effects on DNA synthesis in a defined chromosomal replicon. Mol Cell Biol 1994; 14:1901-1908.
144. Painter RB, Young BR. Radiosensitivity in ataxia-telangiectasia: a new explanation. Proc Natl Acad Sci USA 1980; 77:7315-7317.
145. Chaturvedi P, Eng WK, Zhu Y et al. Mammalian Chk2 is a downstream effector of the ATM-dependent DNA damage checkpoint pathway. Oncogene 1999; 18:4047-4054.
146. Matsuoka S, Huang M, Elledge SJ. Linkage of ATM to cell cycle regulation by the Chk2 protein kinase. Science 1998; 282:1893-1897.
147. Myung K, Datta A, Kolodner RD. Suppression of spontaneous chromosomal rearrangements by S phase checkpoint functions in Saccharomyces cerevisiae. Cell 2001; 104:397-408.
148. D'Amours D, Jackson SP. The yeast Xrs2 complex functions in S phase checkpoint regulation. Genes Dev 2001; 15:2238-2249.
149. Grenon M, Gilbert C, Lowndes NF. Checkpoint activation in response to double-strand breaks requires the Mre11/Rad50/Xrs2 complex. Nat Cell Biol 2001; 3:844-847.
150. Wang H, Elledge SJ. DRC1, DNA replication and checkpoint protein 1, functions with DPB11 to control DNA replication and the S-phase checkpoint in Saccharomyces cerevisiae. Proc Natl Acad Sci USA 1999; 96:3824-3829.
151. Cho RJ, Campbell MJ, Winzeler EA et al. A genome-wide transcriptional analysis of the mitotic cell cycle. Mol Cell 1998; 2:65-73.
152. Imamura O, Fujita K, Shimamoto A et al. Bloom helicase is involved in DNA surveillance in early S phase in vertebrate cells. Oncogene 2001; 20:1143-1151.
153. Somasundaram K. Tumor suppressor p53: regulation and function. Front Biosci 2000; 5:D424-437.
154. Blander G, Kipnis J, Leal JF et al. Physical and functional interaction between p53 and the Werner's syndrome protein. J Biol Chem 1999; 274:29463-29469.
155. Blander G, Zalle N, Leal JF et al. The Werner syndrome protein contributes to induction of p53 by DNA damage. Faseb J 2000; 14:2138-2140.
156. Spillare EA, Robles AI, Wang XW et al. p53-mediated apoptosis is attenuated in Werner syndrome cells. Genes Dev 1999; 13:1355-1360.
157. Wang XW, Tseng A, Ellis NA et al. Functional interaction of p53 and BLM DNA helicase in apoptosis. J Biol Chem 2001; 276:32948-32955.
158. Hodges M, Tissot C, Howe K et al. Structure, organization, and dynamics of promyelocytic leukemia protein nuclear bodies. Am J Hum Genet 1998; 63:297-304.

159. Park JS, Park SJ, Peng X et al. Involvement of DNA-dependent protein kinase in UV-induced replication arrest. J Biol Chem 1999; 274:32520-32527.
160. Frei C, Gasser SM. RecQ-like helicases: The DNA replication checkpoint connection. J Cell Sci 2000; 113:2641-2646.
161. Fujiwara Y, Kano Y, Ichihashi M et al. Abnormal fibroblast aging and DNA replication in the Werner syndrome. Adv Exp Med Biol 1985; 190:459-477.
162. Hanaoka F, Yamada M, Takeuchi F et al. Autoradiographic studies of DNA replication in Werner's syndrome cells. Adv Exp Med Biol 1985; 190:439-457.
163. Ui A, Satoh Y, Onoda F et al. The N-terminal region of Sgs1, which interacts with Top3, is required for complementation of MMS sensitivity and suppression of hyper- recombination in sgs1 disruptants. Mol Genet Genomics 2001; 265:837-850.
164. Kamath-Loeb AS, Loeb LA, Johansson E et al. Interactions between the Werner syndrome helicase and DNA polymerase delta specifically facilitate copying of tetraplex and hairpin structures of the d(CGG)n trinucleotide repeat sequence. J Biol Chem 2001; 276:16439-16446.
165. Harmon FG, Kowalczykowski SC. RecQ helicase, in concert with RecA and SSB proteins, initiates and disrupts DNA recombination. Genes Dev 1998; 12:1134-1144.
166. de Lange T. Protection of mammalian telomeres. Oncogene 2002; 21:532-540.
167. Lowell JE, Pillus L. Telomere tales: Chromatin, telomerase and telomere function in Saccharomyces cerevisiae. Cell Mol Life Sci 1998; 54:32-49.
168. Lundblad V. DNA ends: Maintenance of chromosome termini versus repair of double strand breaks. Mutat Res 2000; 451:227-240.
169. Shay JW, Wright WE. Telomeres and telomerase: Implications for cancer and aging. Radiat Res 2001; 155:188-193.
170. Bodnar AG, Ouellette M, Frolkis M et al. Extension of life-span by introduction of telomerase into normal human cells. Science 1998; 279: 349-352
171. Hahn WC, Stewart SA, Brooks MW et al. Inhibition of telomerase limits the growth of human cancer cells. Nat Med 1999; 5:1164-1170.
172. Henson JD, Neumann AA, Yeager TR et al. Alternative lengthening of telomeres in mammalian cells. Oncogene 2002; 21:598-610.
173. Bryan TM, Marusic L, Bacchetti S et al. The telomere lengthening mechanism in telomerase-negative immortal human cells does not involve the telomerase RNA subunit. Hum Mol Genet 1997; 6:921-926.
174. Bryan TM, Englezou A, Dalla-Pozza L et al. Evidence for an alternative mechanism for maintaining telomere length in human tumors and tumor-derived cell lines. Nat Med 1997; 3:1271-1274.
175. Grobelny JV, Godwin AK, Broccoli D. ALT-associated PML bodies are present in viable cells and are enriched in cells in the G(2)/M phase of the cell cycle. J Cell Sci 2000; 113 Pt 24:4577-4585.
176. Yeager TR, Neumann AA, Englezou A et al. Telomerase-negative immortalized human cells contain a novel type of promyelocytic leukemia (PML) body. Cancer Res 1999; 59:4175-4179.
177. Lendvay TS, Morris DK, Sah J et al. Senescence mutants of Saccharomyces cerevisiae with a defect in telomere replication identify three additional EST genes. Genetics 1996; 144:1399-1412.
178. Lundblad V, Szostak JW. A mutant with a defect in telomere elongation leads to senescence in yeast. Cell 1989; 57:633-643.
179. Lundblad V, Blackburn EH. An alternative pathway for yeast telomere maintenance rescues est1- senescence. Cell 1993; 73:347-360.
180. Kass-Eisler A, Greider CW. Recombination in telomere-length maintenance. Trends Biochem Sci 2000; 25:200-204.
181. Teng SC, Zakian VA. Telomere-telomere recombination is an efficient bypass pathway for telomere maintenance in Saccharomyces cerevisiae. Mol Cell Biol 1999; 19:8083-8093.
182. Le S, Moore JK, Haber JE et al. RAD50 and RAD51 define two pathways that collaborate to maintain telomeres in the absence of telomerase. Genetics 1999; 152:143-152.
183. Kraus E, Leung WY, Haber JE. Break-induced replication: A review and an example in budding yeast. Proc Natl Acad Sci USA 2001; 98:8255-8262.
184. Malkova A, Signon L, Schaefer CB et al. RAD51-independent break-induced replication to repair a broken chromosome depends on a distant enhancer site. Genes Dev 2001; 15:1055-1060.

185. Wyllie FS, Jones CJ, Skinner JW et al. Telomerase prevents the accelerated cell ageing of Werner syndrome fibroblasts. Nat Genet 2000; 24:16-17.
186. Ohsugi I, Tokutake Y, Suzuki N et al. Telomere repeat DNA forms a large non-covalent complex with unique cohesive properties which is dissociated by Werner syndrome DNA helicase in the presence of replication protein A. Nucleic Acids Res 2000; 28:3642-3648.
187. Gravel S, Larrivee M, Labrecque P et al. Yeast Ku as a regulator of chromosomal DNA end structure. Science 1998; 280:741-744.
188. Hsu HL, Gilley D, Blackburn EH et al. Ku is associated with the telomere in mammals. Proc Natl Acad Sci USA 1999; 96:12454-12458.
189. Song K, Jung D, Jung Y et al. Interaction of human Ku70 with TRF2. FEBS Lett 2000; 481:81-85.
190. Shiratori M, Sakamoto S, Suzuki N et al. Detection by epitope-defined monoclonal antibodies of Werner DNA helicases in the nucleoplasm and their upregulation by cell transformation and immortalization. J Cell Biol 1999; 144:1-9.
191. Mohaghegh P, Hickson ID. DNA helicase deficiencies associated with cancer predisposition and premature ageing disorders. Hum Mol Genet 2001; 10:741-746.
192. Meyn MS. Chromosome instability syndromes: Lessons for carcinogenesis. Curr Top Microbiol Immunol 1997; 221:71-148.
193. Oshima J, Campisi J, Tannock TC et al. Regulation of c-fos expression in senescing Werner syndrome fibroblasts differs from that observed in senescing fibroblasts from normal donors. J Cell Physiol 1995; 162:277-283.
194. Hoeijmakers JH. Genome maintenance mechanisms for preventing cancer. Nature 2001; 411:366-374.
195. Robertson KD, Jones PA. DNA methylation: Past, present and future directions. Carcinogenesis 2000; 21:461-467.
196. Smith SS, Crocitto L. DNA methylation in eukaryotic chromosome stability revisited: DNA methyltransferase in the management of DNA conformation space. Mol Carcinog 1999; 26:1-9.
197. Jackson SP. Genomic stability. Silencing and DNA repair connect. Nature 1997; 388:829-830.
198. Fritze CE, Verschueren K, Strich R et al. Direct evidence for SIR2 modulation of chromatin structure in yeast rDNA. Embo J 1997; 16:6495-6509.
199. Sinclair DA, Mills K, Guarente L. Molecular mechanisms of yeast aging. Trends Biochem Sci 1998; 23:131-134.
200. Defossez PA, Prusty R, Kaeberlein M et al. Elimination of replication block protein Fob1 extends the life span of yeast mother cells. Mol Cell 1999; 3:447-455.
201. Kaeberlein M, McVey M, Guarente L. The SIR2/3/4 complex and SIR2 alone promote longevity in Saccharomyces cerevisiae by two different mechanisms. Genes Dev 1999; 13:2570-2580.
202. Lin S, Defossez PA, Guarente L. Requirement of NAD and SIR2 for life-span extension by calorie resrtiction in Saccharomyces cerevisiae. Science 2000; 289:2126-2128.
203. Kim S, Benguria A, Lai CY et al. Modulation of life-span by histone deacetylase genes in Saccharomyces cerevisiae. Mol. Biol. Cell. 1999; 10:3125-3136.
204. Park PU, Defossez PA, Guarente L. Effects of mutations in DNA repair genes on formation of ribosomal DNA circles and life span in Saccharomyces cerevisiae. Mol Cell Biol 1999; 19:3848-3856.
205. Ashrafi K, Lin SS, Manchester JK et al. Sip2p and its partner snf1p kinase affect aging in S. cerevisiae. Genes Dev 2000; 14:1872-1885.
206. Sinclair DA, Mills K, Guarente L. Aging in Saccharomyces cerevisiae. Annu Rev Microbiol 1998; 52:533-560.
207. Defossez PA, Park PU, Guarente L. Vicious circles: A mechanism for yeast aging. Curr Opin Microbiol 1998; 1:707-711.
208. Gasser SM, Cockell MM. The molecular biology of the SIR proteins. Gene 2001; 279:1-16.
209. Kennedy BK, Gotta M, Sinclair DA et al. Redistribution of silencing proteins from telomeres to the nucleolus is associated with extension of life span in S. cerevisiae. Cell 1997; 89:381-391.
210. Smeal T, Claus J, Kennedy B et al. Loss of transcriptional silencing causes sterility in old mother cells of S. cerevisiae. Cell 1996; 84:633-642.
211. Sinclair DA. unpublished result.
212. Sinclair DA. Paradigms and pitfalls of yeast aging research. Mech Ageing Dev 2002; in press.
213. Lu J, Mullen JR, Brill SJ et al. Human homologues of yeast helicase. Nature 1996; 383:678-679.

214. Seki T, Wang WS, Okumura N et al. cDNA cloning of mouse BLM gene, the homologue to human Bloom's syndrome gene, which is highly expressed in the testis at the mRNA level. Biochim Biophys Acta 1998; 1398:377-381.
215. Martin GM, Oshima J, Gray MD et al. What geriatricians should know about the Werner syndrome. J Am Geriatr Soc 1999; 47:1136-1144.

CHAPTER 7

Potential Function of the Werner's Syndrome Homologue in the African Clawed Frog and the Mouse

Michel Lebel and Philip Leder

Abstract

After the discovery of the gene responsible for WS in the human, genes with high homology to *WRN* were found in the mouse *Mus musculus* and the African clawed frog *Xenopus laevis* genome. These laboratory animals brought new experimental approaches in which to study and understand the potential function of the WS gene product. Indeed, these laboratory animals have contributed to our understanding of the potential activity of the WRN protein at the DNA replication fork and its role in DNA recombination/repair. Finally, mouse models of WS were successfully used to identify and study genes that also affected different aspects of aging including cancer.

Introduction

Werner's syndrome (WS) is a rare autosomal recessive disorder characterized by the premature onset of a number of processes associated with aging including malignancies.[1,2] The cause of death is either cancer or cardiovascular disease and occurs at a median age of 47 years.[1] The phenotype of cultured fibroblasts explanted from patients suffering from WS also suggests a parallel between WS and aging. The proliferative life span of WS fibroblasts is reduced compared to age-matched controls. Thus, WS cells behave like fibroblasts established from elderly individuals.[3-5] WS cells are also characterized by an increased genomic instability.[6] The gene responsible for WS (*WRN*) was identified by positional cloning.[7] The *WRN* gene product contains seven helicase consensus domains that are 34 to 38% identical to the *Escherichia coli* RecQ gene,[8] and to the putative yeast helicase Sgs1p.[9,10] The protein also contains a 3'-5' exonuclease activity in addition to its 3'-5' helicase activity.[11-13] Shortly after the discovery of this gene in the human, genes with high homology to *WRN* were found in the mouse *Mus musculus* and the African clawed frog *Xenopus laevis* genomes.[14,15] These laboratory animals brought new experimental approaches in which to study and understand the pathogenesis of WS. Animals such as mice can be genetically manipulated to study specific genes in different disorders. Indeed, animal models of human diseases have been developed which mimicked many aspects of human afflictions, and have allowed novel therapies to be developed. The

Molecular Mechanisms of Werner's Syndrome, edited by Michel Lebel. ©2004 Eurekah.com and Kluwer Academic / Plenum Publishers.

focus of this chapter is to provide an insight into what we have learned so far on the *WRN* gene product using the mouse and the frog systems.

The Mouse and Frog *WRN* Gene Homologues

The amino acid sequence of the human *WRN* gene product revealed two potential catalytic domains. One of them is an helicase domain with high homology to several members of a subfamily of helicases containing a DEXH box sequence.[7] This subfamily includes the RecQ helicase prototype in *Escherichia coli*[8] as well as RecQL, RecQ4, RecQ5, and BLM helicases in human.[16-19] It also includes Sgs1 from the budding yeast *Saccharomyces cerevisiae* and Rqh1 in the fission yeast *Schizosaccharomyces pombe*.[9,10,20] RecQ in *E. coli*, Sgs1 in *S. cerevisiae*, and Rqh1 in *S. pombe* are considered orthologues of the human *WRN* gene rather than homologues. Homology at the amino acid level between these proteins and the WS peptide is high in the helicase domain but poor outside of it. In contrast, the homology between the mouse Wrn peptide or the *Xenopus laevis* FFA-1 gene product and the human *WRN* gene product covers the entire sequence.[14,15,21] FFA-1 or focus forming activity 1 is required for the formation of replication foci in *Xenopus laevis* egg extracts.[22] Like FFA-1, the mouse Wrn protein is also found to be purified with replication complexes in embryonic stem cells.[23] Both the mouse Wrn and the *Xenopus* FFA-1 proteins have a potential exonuclease and a potential helicase catalytic domain.

The FFA-1 gene codes for a protein of 1436 amino acids. The overall similarity between FFA-1 and the human WRN peptide is 66% with an identity of 50%. The middle of the protein (codons 505-802) forms the highly conserved helicase domain. Biochemical analyses of purified FFA-1 from fractionated egg extracts and recombinant FFA-1 produced in bacteria have been performed to characterize this protein.[15,24] These analyses have indicated that FFA-1 contains a 3'-5' exonuclease activity at its N-terminus portion like the human WRN protein, a DNA dependent ATPase activity, and a helicase activity.

The mouse *Wrn* gene codes for a protein of 1401 amino acids. The overall similarity between mouse Wrn and human WRN peptides is 80% with an identity of 72%.[14,25] The helicase domain alone of the mouse Wrn is 95% identical to the human WRN protein.[14] Sequence analysis of the exonuclease domain alone indicates a 82% identity between the mouse and the human sequences. Analyses of the recombinant human WRN and mouse Wrn proteins in vitro have indicated that both peptides have a 3'-5' exonuclease activity with similar DNA substrate specificity.[26] Further characterization of the exonuclease domain indicated that the human WRN and mouse Wrn efficiently degrades the 3' recessed strands of double-stranded DNA or a DNA-RNA heteroduplex. It has little or no activity on blunt-ended DNA duplex, DNA duplex with a 3' protruding strand, or single-stranded DNA. The WRN exonuclease can efficiently remove a mismatched nucleotide at a 3' recessed terminus, and is capable of initiating DNA degradation from a 12-nucleotide gap or a nick.[26] This exonuclease activity is independent of the helicase domain as Wrn peptides containing only the first N-terminal 333 residues retain the exonuclease activity.[11,26] The direction of the helicase activity of the mouse Wrn protein has not been analyzed thoroughly to this date. However, as the homology of the mouse and human WRN helicase domains is greater than 94%, it is likely that the mouse Wrn peptide has 3'-5' helicase activity similar to the human WRN protein.[13,27-29]

Sequence analyses of some of the mouse exon-intron boundaries have revealed that the homology between the human *WRN* and the mouse *Wrn* genes is not only high at the amino acid level, but also at the exon structural level. Sequences of part of the mouse *Wrn* genomic fragments has revealed that the conserved motifs II, III, IV and V of the helicase domain are encoded by separate exons like in the human gene homologue.[30] In addition, exons coding for part of the carboxy terminus portion of the mouse protein have similar exon-intron boundaries.[21,25] More importantly, the mouse *Wrn* gene has been mapped to a region of mouse chromosome 8 syntenic to the human chromosome 8p12 region that contains the human *WRN* gene.[14,21] The syntenic chromosomal location of mouse and human *WRN* genes and their tight sequence homologies strongly support the conclusion that these genes are homologues of one another. In addition, the mouse *Wrn* gene is ubiquitously expressed in all tissues as is the human *WRN* gene.[7,14] Probing of Northern blots at reduced stringency did not reveal any transcript which might correspond to a second *Wrn* gene in mice.[25] Finally, analyses of genomic DNA from different animals (zoo blot) have indicated that there is a *WRN* homologue in several other species as well.[14] Thus, the WRN protein sequence and enzymatic activities have been conserved through evolution.

Differences between Mouse Wrn and Human WRN Proteins

A major difference between the human WRN and mouse Wrn peptides is a fragment of 27 amino acid residues that is repeated twice in the human peptide. The acidic fragment is only found once in mouse or rat.[14] This 27 amino acid repeat is located between the exonuclease and the helicase domains and is encoded by two different exons in the human genome.[30] Careful analysis of the mouse genomic DNA revealed only one exon for this 27-residue fragment (M. Lebel, unpublished results). Interestingly, the *Xenopus* FFA-1 peptide sequence does not contain a duplication of this 27 amino acid region.[15] These observations suggest that the 27 amino acid repeat is a recent duplication that occurred in the human genome during evolution.

Helicases play important roles in a variety of DNA transactions, including DNA replication, transcription, repair, and recombination. In this context, several laboratories have assessed the role of WRN protein in transcription. A potential role of WRN in transcription was first described in yeast one-hybrid system.[31] Soon after this observation, it was found that the WRN peptide could transactivate a number of promoters in vitro and in vivo.[32,33] More importantly, analyses have indicated that the 27 amino acid repeats in the human WRN protein are absolutely required for transactivation.[31,32] Interestingly, a construct containing only one such repeat can transactivate the same promoters.[32] There was no obvious difference between the activity of a WRN peptide with one or two repeats in the context of RNA polymerase II transcription. These results suggest that the mouse Wrn peptide, which contains only one such repeat could transactivate a promoter as well. There would be no major differences in terms of transcriptional activation between the mouse Wrn and the human WRN proteins. However, experiments with the mouse Wrn protein and a specific promoter are still required.

A second major difference between the mouse Wrn and the human WRN protein is their localization in the cells as revealed by immunofluorescence studies. In exponentially growing fibroblasts, the human WRN protein is predominantly localized in the nucleoli.[34,35] In resting cells or upon treatment of the cells with different DNA damaging agents, the WRN protein redistributes itself in the nucleus.[36,37] In contrast, the mouse Wrn protein is distributed throughout the nucleus and not specifically localized to the nucleoli.[35,37,38] Interestingly, localization of the mouse Wrn protein in human cells is nucleoplasmic and not nucleolar.[37] Careful analysis of the WRN proteins have indicated that the inability of the mouse Wrn protein to migrate into the nucleolus is due to a difference in the sequence of the region corresponding to the nucleolar localization signal of the human WRN peptide.[37] The C-terminus of the human WRN peptide contains an arginine-lysine motif required for the nucleolar localization. This motif is absent in the mouse Wrn protein. Furthermore, mouse cells do not recognize this arginine-lysine motif of human WRN protein and do not translocate the human protein in the nucleoli implying a different nucleolar trafficking system between mouse and human cells.[37] The exact function of the WRN protein in human nucleoli is still unknown as several human cell lines do not show such nucleolar localization.[38] Despite the differences between mouse Wrn and human WRN proteins, studies on the laboratory mouse have provided important clues to the possible function of WRN in cells.

Function of the WRN Homologue in the *Xenopus* System

The first clue to the potential function of WRN in cells came from in vitro studies of the replication foci in *Xenopus laevis* egg extracts. Extracts derived from unfertilized eggs offer a good model system which allows biochemical approaches in addition to visual observations. In this system, introduction of demembranated sperm chromatin will induce nuclear envelope formation around itself and the DNA will be replicated once semiconservatively. As in somatic cells, replication in these reconstituted nuclei occurs at a large number of discrete foci.[39] In addition, structures similar to replication foci can also form on chromatin, even in the absence of membrane.[22] One component of these foci is the single-stranded DNA binding protein RPA (replication protein A), which is essential for eukaryotic DNA replication.[41] Fractionation experiments have indicated that RPA associates with foci on chromatin only in the presence of another protein, focus forming activity 1.[40] Sequence analysis of the gene encoding FFA-1 has revealed an extensive homology between FFA-1 and the human *WRN* gene product throughout the open reading frame, suggesting that FFA-1 is the first true homologue of human WRN outside mammals.[15] Thorough analyses of the replication foci in reconstituted egg extracts have indicated that FFA-1 colocalizes specifically with sites of DNA synthesis.[24] RPA was shown to bind and stimulate the DNA helicase activity of FFA-1 at sites of DNA synthesis. These results provided the first evidence of an important role for FFA-1 in DNA replication.[24] Interestingly, however, immunodepletion of FFA-1 in reconstituted egg extracts did not completely abolish DNA replication in nuclei. These results indicated that FFA-1 is important but not essential for DNA replication.[24] Such a conclusion also applies to the human WRN protein. It has been shown that the human WRN protein associates with several proteins found at the DNA replication fork. It can bind the replication protein A (RPA), the proliferation cell nuclear antigen (PCNA), topoisomerase I and more importantly the small subunit of DNA polymerase delta (p50).[13,23,26,42,43] The exact function of WRN protein at the replication fork is still unknown. Like FFA-1, WRN protein is not absolutely required for DNA replication as DNA synthesis is detected in Werner's syndrome cells.[44-46]

One advantage of using the reconstituted *Xenopus* egg extracts on sperm chromatin, is the possibility of adding different recombinant fragments of the WRN peptide homologue FFA-1, without having to worry about the presence of an intact nuclear localization signal. The nuclear localization signal was identified near the end of the carboxy terminus of the WRN protein.[47,48] Such reconstitution experiments have demonstrated that RPA binds to a region of FFA-1 located between the exonuclease and the helicase domains.[24] This fragment of FFA-1 can localize to the replication foci with the intact endogenous FFA-1 as visualized by immunofluorescence.[24] Interestingly, this fragment of FFA-1 has a dominant negative effect as little DNA synthesis was detected at the replication foci.[24] A fragment of FFA-1 containing only the exonuclease domain did not have a dominant negative effect even though the activity of this exonuclease is high in vitro. It has been suggested that the dominant negative FFA-1 recombinant fragment prevents stimulation of helicase activities of intact endogenous FFA-1 by RPA at replication foci.[24] Thus, these experiments demonstrate how powerful the reconstituted egg extract is to visualize and study the activity of FFA-1 at replication foci in eukaryotic cells. Using this system, it will be possible to specifically inactivate the helicase, the exonuclease, or both domains by point mutations on recombinant FFA-1 proteins and then determine if such mutated proteins affect DNA synthesis at replication foci in FFA-1 immunodepleted egg extracts. Specific point mutations inactivating either the helicase or exonuclease domains have been described for the human WRN protein.[13,26] Reconstitution experiments with wild type or mutated recombinant FFA-1 should tell us if both the exonuclease and the helicase domains are required for DNA replication.

Function of the Wrn Homologue in Mouse

Werner's syndrome (WS) apparently results from a single gene defect. Although there is variability in the phenotype from one patient to another, this syndrome is characterized by the early onset of osteoporosis, diabetes mellitus, ocular cataracts, early graying of the hair, atherosclerosis, and several types of neoplasms including rare types of cancer.[49] WS also includes features not associated with aging such as short stature, hyperkeratosis, subcutaneous atrophy, trophic ulcers of the legs, telangiectasia, increased hyaluronic acid in the urine and hypogonadism. Therefore, WS is phenotypically distinct from normal aging and is viewed more as a caricature of accelerated aging.[50] At the cellular levels, WS cells exhibit variegated chromosomal translocations and deletions. Such chromosomal abnormalities were found in vitro and in vivo in skin fibroblast cell lines as well as from lymphoblastoid cell lines made from circulating lymphocytes of WS patients.[3,51,52] As metioned earlier, the WRN gene product shows similarity to DNA helicases of the RecQ subfamily[7] and contains an exonuclease domain with homologies to the exodeoxyribonuclease proofreading domain of *E. coli* DNA polymerase I and ribonuclease D.[53] DNA helicases and exonucleases perform many important functions in replication, transcription, recombination, and DNA repair. It is currently not clear how such enzymes lead to premature aging. However, inactivation of these proteins may lead to defects in important events such as senescence, apoptosis, and cell transformation. The structural similarities of mouse and human proteins suggest that mice could provide a model or a biological tool to study different aspects of the disease. For this reason, several laboratories created mice with different mutations in the murine *Wrn* gene.

The first WS mouse model was created by deleting part of the helicase domain of the murine *Wrn* in embryonic stem cells.[21] The two exons encoding motifs III and IV of the helicase domain were deleted by homologous recombination. Such mutation created an in-frame deletion in the *Wrn* gene. Deletion of part of the helicase domain has also been described for some human WS patients.[30] However, no stable WRN protein was detected in such patients unlike this mouse model which synthesizes a stable mutant protein in vivo.[21] In fact, no WRN protein can be detected in any of the WS patients with different mutations.[54] This is a major difference between this mouse model and all WS patients. Nevertheless, analyses of cells derived from this mouse model corroborate many aspects of human WS cells as will be discussed below. Moreover, with this mouse model, the role of the helicase domain in cells can be addressed specifically.

The second knock out model of the murine *Wrn* gene described in the literature, consists of a mutation that eliminates the expression of the carboxy terminus region of *Wrn*.[25] This mouse model turns out to be a *Wrn* null mutant as no Wrn protein was detected using antibodies against either the amino terminus or the carboxy terminus part of the protein.

A third mouse model of WS described in the literature was generated by introducing a human cDNA with a dominant negative mutation in the helicase domain into mice.[55]

Cellular Phenotype of *Wrn* Mutant Mice

One important characteristic of mice with an in-frame deletion of the helicase domain is that the stable mutant Wrn protein can be detected in different fractions of each fractionation by Western blot analyses. Using the cells from this mouse model, it was found that wild type, but not mutant Wrn protein, copurifies through a series of centrifugation, chromatography, and sucrose gradient steps with the well-characterized 17S multiprotein DNA replication complex.[23] Furthermore, wild type WS protein coimmunoprecipitates with a prominent component of the multiprotein replication complex, proliferating cell nuclear antigen (PCNA). In vitro studies have also indicated that PCNA binds to a region in the N-terminus portion of the Wrn protein near the 3'-5' exonuclease domain. Following these results, coimmunoprecipitation experiments have confirmed the interaction of not only the mouse Wrn, but also the human WRN protein with several components of the DNA replication fork.[13,23,26,42,43]

Thus, genetic and molecular analyses of this mouse model provided the first physical evidence for a role of the Wrn protein in the DNA replication complex in mammalian cells. At the time, the data made perfectly good sense as WS cells from a human patient demonstrated an impaired S phase transit.[46] Moreover, the rate of initiation of DNA replication was found to be retarded in WS cells compared to control cells.[44,45] The fractionation procedure also indicated that not all murine Wrn proteins in cells are associated with the DNA replication complex.[21,23] The Wrn protein is associated with other complexes in the cells that have not been identified yet. These complexes could be involved in other aspects of DNA metabolism. Indeed, WRN is believed to be involved in DNA recombination/repair and/or transcription.[32,56]

Cell lines derived from mice with a deletion in the helicase domain have been shown to be defective in their response to agents that perturb DNA replication.[21,55] Like human WS cells, mouse embryonic stem cells with a deletion of the helicase domain are sensitive to both type I and II topoisomerase inhibitors, camptothecin and etoposide, respectively.[57-59] Interestingly, heterozygous mutant cells, which express half the amount of intact Wrn protein, had an intermediate sensitivity to these drugs compared to wild type or homozygous mutant cells.[21] Type I topoisomerase (which nicks DNA creating a single-strand break) is involved in some aspects of replication and transcription. Mammalian type II topoisomerase (which cleaves both strands of DNA)[60] is involved in the terminal stages of DNA replication. It is also an important structural component of the mitotic chromosomal scaffold.[61] Camptothecin and etoposide stabilize the DNA/topoisomerase cleavage complex, resulting in DNA strand breaks that may not be religated or repaired. The effect of topoisomerase inhibitors suggests that the WRN helicase acts in concert with topoisomerases and this potential interaction could take place during DNA replication. Indeed, topoisomerase I was coimmunoprecipitated with WRN protein from human cell lysates.[23]

Another characteristic of embryonic stem cells with a deletion of the helicase domain is the increased genomic instability detected at the X-linked enzyme hypoxanthine-guanine phosphoribosyl-transferase (*HPRT*) locus.[21] The anti-leukemic agent 6-thioguanine is a purine analogue which induces its lethal effect as a result of its incorporation into DNA.[62,63] However, in order for the 6-thioguanine to be incorporated into DNA, it requires the action of the enzyme hypoxanthine-guanine phosphoribosyl-transferase. Any spontaneous mutation inactivating the *HPRT* locus will render cells resistant to 6-thioguanine. It was found that the average frequency of 6-thioguanine-resistant colonies was higher in homozygous mutant *Wrn* embryonic stem cells than in either wild type or heterozygous cells.[21] Southern analyses of the resistant colonies have indicated the presence of rearrangements (mainly deletions) at the *HPRT* gene (M. Lebel, unpublished results). Such deletions were also described for 6-thioguanine-resistant human WS cell lines.[64-66] It has been suggested from such studies that the deletion events resulted from illegitimate recombination between small regions of homology within the *HPRT* locus. Interestingly, Southern analyses on wild type resistant mouse embryonic stem cells have revealed similar types of illegitimate recombination leading to deletions within the *HPRT* locus (M. Lebel, unpublished results). These results suggest that the mechanism leading to illegitimate recombination at the *HPRT* locus is the same in wild type and *Wrn* mutant cells. However, the frequency of such illegitimate recombination is much higher in the absence of a functional Wrn helicase. Although more experiments are required on these mouse cells, studies on human WS cells and the WRN peptide have suggested that the WRN protein is a suppressor of illegitimate recombination.[67]

Tail fibroblasts derived from transgenic mice bearing a dominant negative human WRN protein have been shown to be sensitive to the pro-carcinogen 4-nitroquinoline 1-oxide (4-NQO).[55] This agent will cause alkylation of the DNA and induce oxidative stress in cells.[68] In addition, tail fibroblasts expressing the dominant negative human WRN protein also have a reduced replicative life span. These results corroborate the data obtained with human WS cells.[36,69] The sensitivity to 4-NQO and the reduced replicative life span is not seen in fibroblasts from tails of mice expressing a wild type human WRN protein.[55] The mutation in the human WRN protein is a point mutation inactivating the ATPase and helicase activity of the peptide.[13,55] It is believed that this mutant helicase protein associates with protein complexes in mouse cells affecting their normal activities. Once again, this mouse model of WS illustrates the important role the Wrn helicase has in cells. Transgenic or knock out mice with a mutation only in the exonuclease domain are required to assess the effect of this domain on the replicative life span and sensitivity of cells to specific DNA damaging drugs.

Embryonic fibroblasts from homozygous mice that are null for the *Wrn* gene also have a reduced replicative life span.[25] However, there are discrepancies between the sensitivity of these *Wrn* null mouse embryonic fibroblasts to specific DNA damaging agents and cells from the two other WS mouse models discussed above. *Wrn* null cells did not show an increased sensitivity to 4-NQO or to the topoisomerase I inhibitor camptothecin.[25] The exact reason for this discrepancy in the data is unknown. It could be due to the origin of the cells used in each study. The extent of sensitivity of *Wrn* mutant cells to topoisomerase inhibitors is better seen with very rapidly dividing embryonic stem cells[21] than with fibroblasts. Cycling embryonic stem cells have a very short G1 phase and are essentially in S phase. Topoisomerase inhibitors like camptothecin are very potent drugs in S phase of the cell cycle.[70] The discrepancy in the drug sensitivity data may also be due to the fact that studies on the *Wrn* null mice were done with mouse fibroblasts that had another mutation in another RecQ type helicase involved in Bloom's syndrome (BLM in human).[25] The exact effect of DNA damaging drugs has not been described for *Blm* heterozygous or homozygous mutant murine cell lines. In addition, the experiments were performed with homozygous and heterozygous *Wrn* mutant cells. It is now well established that heterozygous WS mutant cells in both human and mouse models are more sensitive to topoisomerase inhibitors or 4-NQO than wild type cells.[21,36] Hence, additional comparison analyses have to be performed on wild type mouse embryonic fibroblasts and fibroblasts that are null only for the murine *Wrn* gene.

Phenotype of *Wrn* Mutant Mice

Careful pathological and pathophysiologic studies have only been reported for aging cohorts of mice with a deletion of the helicase domain of the Wrn protein.[71] All homozygous *Wrn* mutant mice died or were sacrificed because of illnesses by the 25th month of age. Approximately 20% of the heterozygous animals were free of any abnormal symptoms or behavior by this time. In contrast, more than 60% of wild type littermates were still healthy by the age of 25 months. The median age of illness in the homozygous mice was reduced to 20 months of age compared to 23 months in the heterozygotes. More than three quarters of the wild type animals did not develop any obvious phenotype by 20 months of age (Fig. 1). These results indicate that homozygous mice with a deletion of part of the helicase domain of the Wrn protein develop symptoms sooner than wild type littermates[71]

Thus, homozygous mice appeared sick at a much earlier age than wild type animals. This was so even though all animals (wild type, heterozygous, and homozygous) were housed under the same conditions and often in the same cages. The genetic background of these mice is from 129/SvEv and Black Swiss strains. Black Swiss outbred strain is derived from C57BL/6 mice (Taconic, Germantown, NY). Such a strain develops mainly lymphomas at a mean age of 27 months.[72] The few wild type animals that were sick in the colony had either myeloid dysplasia

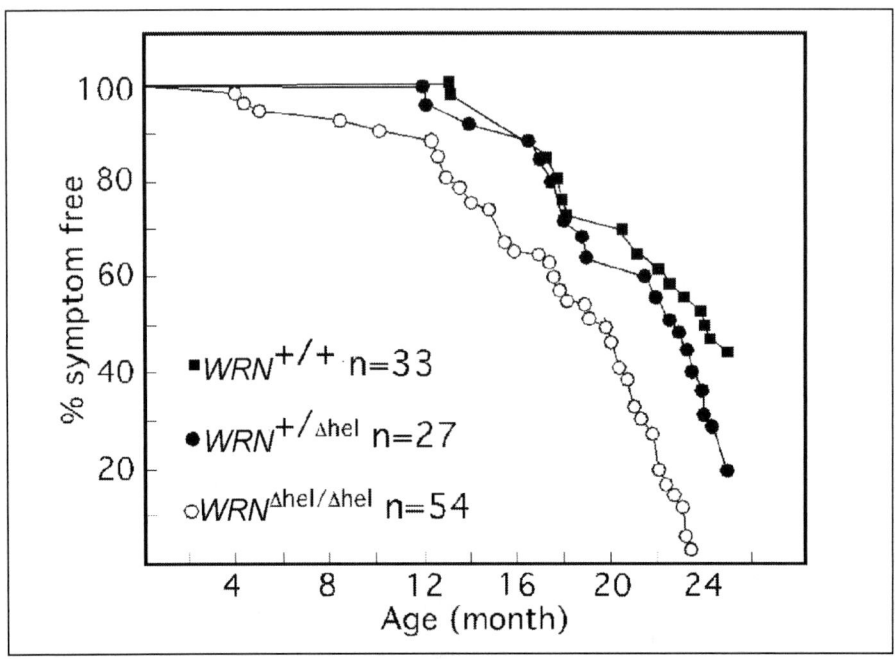

Figure 1. Percentage of symptom free animals for wild type, heterozygous and homozygous *Wrn* mutant mice over a period of two years. Animals were checked three times a week for any external mass, infection, bleeding, gasping, and overall decrease or change in activity or behavior. The dates of such observations were recorded and considered the first apparition of a symptom in each mouse. Mice were kept for additional two to four weeks to assess any decline in the health status and for pathologic examinations. The numbers of animal (n) in this survey are indicated for each genotype.

with an enlarged spleen, infection of the lungs, or mild cardiac fibrosis. The homozygous mice were remarkable with respect to the variety of tumors and apparent inflammations of different organs.[71] Interestingly, homozygous mice also developed unusual types of tumors for this strain such as mammary kerathocanthomas, serous ovarian cystadenoma, ovarian granuloma cell tumors, granulocytic sarcomas, or lymphoblastic lymphomas. In addition to the tumors, several mice also had developed infections or inflammations associated with either the lungs, the perivascular system, or the urinary tract demonstrating multiple abnormal conditions. Several homozygous mice died without obvious illness and tissue autolysis prevented a thorough pathological examination. Heterozygous mice developed different hyperplasias and benign tumors. However, the phenotype in the homozygous animals was more severe than the heterozygous mice and this was independent of the genetic background.

Approximately 60% of homozygous mice developed mild to severe cardiac fibrosis.[71] Few wild type animals showed a mild cardiac fibrosis before the age of 24 months. Hearts from homozygous mice have shown moderate to severe interstitial fibrosis often in the left ventricle with atrophic scar at the apex. Fibrosis was also present around the aortic root or was prominent around the penetrating coronary artery. In some animals one of the auricular appendages had partially organized thrombi implying that those mice had arrhythmia (M. Lebel, unpublished data). Thus, all these results indicate that homozygous *Wrn* mice develop severe cardiac fibrosis at an earlier age than wild type mice. Interestingly, the expression of *Wrn* mRNA is known to be increased in the heart of aging mice.[73] Similarly, the expression of *Wrn* specific mRNA is increased in the liver and the pancreas of 24 month-old mice compared to 2 month-old mice.[73] Thus, the requirement for a higher expression of *Wrn* gene in these murine organs maybe important for normal tissue homeostasis. In the absence of an intact Wrn helicase, mice can develop cardiac fibrosis or tumors in the liver or the pancreas.[71] Heterozygous animals can develop similar diseases, although at a later onset than homozygous animals (Fig. 1). These observations strongly suggest that a deleterious phenotype associated with the heterozygotes can be of potential health concern. Indeed, it has been suggested that human heterozygous carriers may have an increased chance of developing cardiac arrest or cancer.[36,74]

Although the phenotype seen in *Wrn* mutant mice appears late in the life of these animals (at the age of two years), these mice can develop different types of tumors or die from cardiac failure as in human WS patients.[71] However, there are major differences between mice and human individuals with a mutation in the *WRN* gene. Unlike the mouse, the expression of the *WRN* specific mRNA does not increase in human aortic tissues.[75] The heart phenotype detected in the mouse model for WS is similar to a cardiomyopathy. Although cardiac fibrosis or cardiomyopathy can be of concern in the aging human population,[76] WS patients tend to develop myocardial infarctions instead.[1,77-79]

Another major difference is that *Wrn* mutant mice do not develop symptoms such as osteoporosis, diabetes mellitus, ocular cataracts, early graying of the hair and atherosclerosis. The *Wrn* mutant mice may not live long enough to develop these other symptoms. Alternatively, the exact function of the Wrn protein may differ in the tissues of different species.[80] Finally, we cannot rule out the possibility that the mutant Wrn protein can interact and disrupt the function of specific protein complexes in the cells. Long term studies are required with the Wrn null mice to correlate the data. Nevertheless, mice with a deletion of part of the Wrn helicase domain can be used to study genomic instability and tumor progression in vivo.

Wrn Mutant Mice and the *p53* Tumor Suppressor Gene

One advantage of having a mouse model for WS is that appropriate crosses can be performed to ask specific genetic questions. Two laboratories have demonstrated that the C-terminus portion of p53 binds to a C-terminus portion of the WRN peptide.[33,81] It has been shown that human WS fibroblasts have an attenuated p53-mediated apoptotic response, and this deficiency can be rescued by expression of wild type WRN peptide.[81] An attenuation of a p53-apoptotic response in already genetically unstable WS cells would have implications on cancer progression. Consistent with this hypothesis, *p53* null/*Wrn* mutant mice show an increased mortality rate compared to *p53* null or *Wrn* null mice alone.[25,71] The *p53* null/*Wrn* mutant mice were particularly remarkable in terms of the variety of tumors they developed compared to those that appeared in *p53* null mice. In addition, 34% of *p53* null/*Wrn* mutant mice developed multiple hemangiosarcomas in different organs. In contrast, less than 11% of the *p53* null mice had more than one focus of hemangiosarcoma. Moreover, 34% of the *p53* null/*Wrn* mutant mice simultaneously developed multiple types of tumors. None of the *p53* null mice in this survey developed multiple tumors.[71] These observations indicate that *p53* null/*Wrn* mutant mice rapidly develop aggressive tumors as well as unusual types of tumors compared to *p53* null or *Wrn mutant* mice alone.

Mutations in the p53 gene have been associated with a wide variety of human cancers.[82] Mice deficient in p53 develop normally, but are susceptible to spontaneous tumors. More than 70% of mice develop tumors, especially lymphomas and sarcomas, by 6 months of age.[72,83] p53 is a transcription factor[84] that is increased in response to genotoxic stress such as DNA damage. This increase in p53 protein level is thought to result in transcription of target genes that mediate many functions. Among its transcriptional targets is $p21^{WAF1/CIP1}$, an inhibitor of cyclin-dependent kinase complexes (Cdks),[85,86] PCNA,[87] and DNA synthesis.[88] p53 function is required for G1 arrest[89] and it is believed that this arrest allows cells time to repair DNA damage before being fixed as mutations. Indeed, p53-deficient mice are extremely susceptible to radiation-induced tumorigenesis.[90] The p21 protein is also involved in cell cycle arrest.[91] It is induced by DNA damage and is found associated with inactive cyclin E-CDK2 complexes which are essential for G1 to S phase transition.[92] Furthermore, p21 may slow down DNA synthesis by binding to PCNA to facilitate repair processes. In addition to cell cycle arrest, the ability of p53 to induce apoptosis is thought to be an important factor for its tumor suppressor function.[93] Thus, cells lacking p53 function continue to proliferate, perpetuating potentially oncogenic mutations. It has been reported that *p21* null/ *Wrn* mutant mice do not undergo accelerated tumorigenesis.[71] This suggests that the G1/S checkpoint is not a focus of p53-dependent tumor suppression at least in *Wrn* mutant mice. More likely, the key suppressor pathway in the *Wrn* mouse model is via the induction of apoptosis which is also dependent on p53.

Thus, the *p53* null/ *Wrn* mutant genetic crosses have indicated that the absence of a functional Wrn protein will promote tumorigenesis in tissues containing uncontrolled and rapidly dividing cells that are already genetically unstable. Mice with a mutation of the *Wrn* gene will also develop tumors but much later in life indicating that other mutations are required for the appearance of the tumors. Apparently a nonfunctional *Wrn* protein will not initiate hemangiosarcomas, since *Wrn* mutant mice do not develop such tumors. However, mutation in the *Wrn* gene accelerated the progression and the aggressivity of such tumors in *p53* null mice.[71]

It is now well accepted that cancer arises from the accumulation of mutations, some of which affect cell cycle regulation.[94,95] Consequently, the observation that age-related carcinogenesis has been shown to be associated with the accumulation of multiple genetic abnormalities is not an unexpected finding.[96] It is interesting to note that a region of human chromosome 8 containing the *WRN* gene is often deleted in late stage cancers derived from colon, lung, liver, breast and prostate.[97] Although in most cases the deleted region spans several genes, there is one report of a small deletion spanning only specific exons of the *WRN* gene in metastatic cells from a prostate cancer patient.[97] Thus, mutation in the *WRN* gene can also be found in sporadic cancer. This observation also suggests that inactivation of the *WRN* gene may occur late during the progression of cancer. The increased genomic instability seen in WS cells will certainly affect the rate or the rapidity at which abnormally growing cells accumulate other mutations. This, in turn, will give rise to a more aggressive phenotype as seen in the mouse models described above. More studies are required to determine the incidence of *WRN* mutation in early and late stage human carcinogenesis.

Illegitimate Recombination in Tissues of *Wrn* Mutant Mice

To gain information on the type of genomic DNA that can be mutated in WS cells in vivo and the frequency of such mutations, mice with a deletion of the Wrn helicase domain were crossed to the pink-eyed unstable line and the TG.AC mouse line. The pink-eyed mice (Jackson Laboratory) have an unstable duplication of the *p* gene.[98] The *p* gene normally encodes a melanosomal integral membrane protein responsible for the assembly of a high molecular weight melanin complex that produces the black coat color of wild type mice. Phenotypically, the *pun* mutation results in a dilute, light gray coat color and pink eye color. Deletion events that remove one copy of the duplication lead to a reversion of the *pun* mutation to wild type *p* gene. One deletion event occurring in a premelanocyte in the embryo will cause a visible black spot on the gray fur of offspring after amplification of the premelanocytes. The loss of the duplicated region of the *p* gene and subsequent reversion to wild type may be caused by intrachromosomal recombination event. Spontaneous reversion of a *pun* allele leads to spots on the fur of 2-11% of the offspring.[98,99] The frequency of reversion (or deletion at the *p* locus), as detected by the number of black spots on the fur of the animals, is elevated in mice with a *Wrn* mutant background.[100] Interestingly, this frequency is also significantly elevated in heterozygous *Wrn* mutant mice compared to mice with both *Wrn* wild type alleles. This was the first experiment providing clear evidence for a phenotype in heterozygous individuals at the tissue level.[100] Although the frequency of mutation in heterozygotes is less than in homozygotes, these results strongly suggest that a deleterious phenotype might be associated with the heterozygotes that could be of potential health concern in the general human population.[36,74]

Genetic instability has also been well studied in the TG.AC transgenic mice. The TG.AC transgenic mouse bears a multicopy fusion transgene in which the mouse embryonic ζ-globin promoter drives the expression of the *v-Ha-Ras* oncogene.[101] It turns out that the transgene is inserted into a Line-1 element and the resultant structure creates an extended inverted repeat consisting of both transgene and Line-1 sequences.[102] Interestingly, this multicopy transgene is unstable as rearrangements and deletions have been detected at the transgene locus.[103-105] It has been shown that the rate of rearrangements at this transgenic locus is much higher on a *Wrn* mutant genetic background.[102] In some cases, the deletions are so extensive that not even a single intact copy of the transgene remains.[102] Moreover, such rearrangements have been found not only in somatic cells of the TG.AC/*Wrn* mutant mice, but also in germ cells.[102] Thus, complex repeated DNA sequences or DNA structures in tandem can be unstable in mice. On a *Wrn* mutant genetic background, this DNA instability is much greater. This is an important observation as repeated sequences like telomeres or ribosomal genes could potentially be predisposed to rearrangements in the absence of a functional Wrn protein and could affect cell growth. Indeed, it has been shown that shortened telomeres and rearrangements of ribosomal DNA structures can affect cell senescence in yeast cells.[106-110] Repetitive sequences like telomeres and ribosomal DNA sequences will have to be monitored carefully in these mouse models.

Future Perspectives

A great deal has been learned of the potential function of the WRN protein from frogs and mice. More information will be added with time since it is relatively easy to genetically manipulate these animal models. As discussed in the preceding sections, experiments on frogs and mice have complemented a lot of the research conducted on the human population. In some instances, the data on laboratory animals have contributed to our understanding of the potential function of the WRN protein at the DNA replication fork and its role in DNA recombination/repair. In addition, the mouse models of WS can be used to identify and study new genes that may also affect different aspects of aging. Several proteins have already been identified as forming complexes with the WRN peptide in vitro or in vivo. Such proteins have also been disrupted in the mouse. The best examples are p53, topoisomerase I and the Ku complex.[23,33,81,111,112] The genomic instability and the tumor progression in *p53* null mice is well documented.[72,83] The disruption of the topoisomerase I enzyme in mice causes embryonic lethality at the eight-cell stage.[113] This demonstrates the importance of topoisomerase I in cell division. The Ku86/70 complex, which binds to the ends of double-strand breaks, interacts and stimulates the exonuclease (but not the helicase) activity of the WRN protein again implicating WRN in DNA repair.[111,112] Disruption of the *Ku86* gene in mice is quite striking as mutant mice prematurely exhibited age-specific changes characteristic of senescence that include osteopenia, atrophic skin, hepatocellular degeneration, hepatocellular inclusions, hepatic hyperplastic foci, and age-specific mortality including cancer.[114] Mice with mutations in the exonuclease domain of the Wrn protein will be required to determine the effect of this activity on aging. The *Wrn* mutant mouse models may not be appropriate for the investigations of all the pathological processes leading to WS in human. However, disruption of murine proteins known to genetically interact with Wrn will give us more information regarding different aspects of many age-related diseases.

The mouse models for WS are also important disease surrogates that can be used to analyze the physiological effects of anti-cancer drugs. Studies on the topoisomerase I inhibitor, camptothecin, have clearly demonstrated the increased deleterious effect of this drug on animals with a mutation in the *Wrn* gene.[100] Thus, these mice can not only be genetically manipulated, but they can also be used to study the effect of environmental factors or chemical agents on genetic instability in a mammalian system. The *Wrn* mutant genetic crosses indicated that absence of a functional Wrn protein promotes tumorigenesis in tissues containing rapidly dividing cells that are already genetically unstable. Additional studies are required to determine the impact of the *Wrn* mutation in early and late stage carcinogenesis in our mouse models and in human cancers. Such studies may provide one of early examples whereby therapy will be related to the precise genetic make up (haplotype) of the patient.

Acknowledgements

Michel Lebel is supported by a grant from the Canadian Institutes of Health Research.

References

1. Epstein CJ, Martin GM, Schultz AL et al. Werner's syndrome: A review of its symptomatology, natural history, pathologic features, genetics and relationship to the natural aging process. Medecine 1966; 45:177-222.
2. Sato K, Goto M, Nishioka K et al. Werner's syndrome associated with malignancies: Five case reports with a survey of case histories in Japan. Gerontology 1988; 34:212-218.
3. Salk D, Bryant E, Au K et al. Systematic growth studies, cocultivation, and cell hybridization studies of Werner syndrome cultured skin fibroblasts. Hum Genet 1981; 58:310-316.
4. Faragher RGA, Kill IR, Hunter JAA et al. The gene responsible for Werner syndrome may be a "counting" gene. Proc Nat Acad Sci USA 1993; 90:12030-12034.
5. Schulz VP, Zakian VA, Ogburn CE et al. Accelerated loss of telomeric repeats may not explain accelerated replicative decline of Werner syndrome cells. Hum Genet 1996; 97:750-754.
6. Salk D. Werner's syndrome: A review of recent research with an analysis of connective tissue metabolism, growth control of cultured cells, and chromosomal aberrations. Hum Genet 1982; 62:1-5.
7. Yu C-E, Oshima J, Fu Y-H et al. Positional cloning of the Werner's syndrome gene. Science 1996; 272:258-262.
8. Umezu K, Nakayama K, Nakayama H. Escherichia coli RecQ protein is a DNA helicase. Proc Nat Acad Sci USA 1990; 87:5363-5367.
9. Gangloff S, McDonald JP, Bendixen C et al. The yeast type I topoisomerase Top3 interacts with Sgs1, a DNA helicase homologue: A potential eukaryotic reverse gyrase. Mol Cell Biol 1994; 14:8391-8398.
10. Watt PM, Louis EJ, Borts RH et al. Sgs1: A eukaryotic homolog of E. coli RecQ that interacts with topoisomerase II in vivo and is required for faithfull chromosome segregation Cell 1995; 81:253-260.
11. Huang S, Li B, Gray MD et al. The premature ageing syndrome protein, WRN, is a 3'—>5' exonuclease. Nat Genet 1998; 20:114-116.
12. Kamath-Loeb AS, Shen JC, Loeb LA et al. Werner syndrome protein. II Characterization of the integral 3'—>5' DNA exonuclease J Biol Chem 1998; 273:34145-50.
13. Shen JC, Gray MD, Oshima J et al. Werner syndrome protein. I DNA helicase and dna exonuclease reside on the same polypeptide J Biol Chem 1998; 273:34139-34144.
14. Imamura O, Ichikawa K, Yamabe Y et al. Cloning of a mouse homologue of the human Werner syndrome gene and assignment to 8A4 by fluorescence in situ hybridization. Genomics 1997; 41:298-300.
15. Yan H, Chen CY Kobayashi R et al. Replication focus-forming activity 1 and the Werner syndrome gene product. Nat Genet 1998; 19:375-378.
16. Puranam KL, Blackshear PJ. Cloning and characterization of RECQL, a potential human homologue of the Escherichia coli DNA helicase RecQ. J Biol Chem 1994; 269:29838-29845.

17. Seki M, Miyazawa H, Tada S et al. Molecular cloning of cDNA encoding human DNA helicase Q1 which has homology to Escherichia coli Rec Q helicase and localization of the gene at chromosome 12p12. Nucleic Acids Res 1994; 22:4566-4573.
18. Ellis NA, Groden J, Ye TZ et al. The Bloom's syndrome gene product is homologous to RecQ helicases. 1995; 83:655-666.
19. Kitao S, Ohsugi I, Ichikawa K et al. Cloning of two new human helicase genes of the RecQ family: biological significance of multiple species in higher eukaryotes. Genomics 1998; 54:443-452.
20. Stewart E, Chapman CR, Al-Khodairy F et al. Rqh1+, a fission yeast gene related to the Bloom's and Werner's syndrome genes, is required for reversible S phase arrest. EMBO J 1997; 16:2682-2692.
21. Lebel M, Leder P. A deletion within the murine werner syndrome helicase induces sensitivity to inhibitors of topoisomerase and loss of cellular proliferative capacity. Proc Natl Acad Sci USA 1998; 95:13097-13102.
22. Yan H, Newport J. FFA-1, a protein that promotes the formation of replication centers within nuclei. Science 1995; 269:1883-1885.
23. Lebel M, Spillare EA, Harris CC et al. The Werner syndrome gene product copurifies with the DNA replication complex and interacts with PCNA and topoisomerase I. J Biol Chem 1999; 274:37795-37799.
24. Chen CY, Graham J, Yan H. Evidence for a replication function of FFA-1, the Xenopus orthologue of Werner syndrome protein. J Cell Biol 2001; 152:985-996.
25. Lombard DB, Beard C, Johnson B et al. Mutations in the WRN gene in mice accelerate mortality in a p53-null background. Mol Cell Biol 2000; 20:3286-3291.
26. Huang S, Beresten S, Li B et al. Characterization of the human and mouse WRN 3'—>5' exonuclease. Nucleic Acids Res 2000; 28:2396-2405.
27. Gray MD, Shen JC, Kamath-Loeb AS et al. The Werner syndrome protein is a DNA helicase. Nat Genet 1997; 17:100-103.
28. Suzuki N, Shimamoto A, Imamura O et al. DNA helicase activity in Werner's syndrome gene product synthesized in a baculovirus system. Nucleic Acids Res 1997; 25:2973-2978.
29. Fry M, Loeb LA. Human werner syndrome DNA helicase unwinds tetrahelical structures of the fragile X syndrome repeat sequence d(CGG)n. J Biol Chem 1999; 274:12797-12802.
30. Yu C-E, Oshima J, Wijsman EM et al. Mutations in the consensus helicase domains of the Werner Syndrome gene. Am J Hum Genet 1997; 60:330-341.
31. Ye L, Nakura J, Morishima A et al. Transcriptional activation by the Werner syndrome gene product in yeast. Exp Gerontol 1998; 33:805-812.
32. Balajee AS, Machwe A, May A et al. The Werner syndrome protein is involved in RNA polymerase II transcription. Mol Biol Cell 1999; 10:2655-2668.
33. Blander G, Kipnis J, Leal JF et al. Physical and functional interaction between p53 and the Werner's syndrome protein. J Biol Chem 1999; 274:29463-29469.
34. Gray MD, Wang L, Youssoufian H et al. Werner helicase is localized to transcriptionally active nucleoli of cycling cells. Exp Cell Res 1998; 242:487-494.
35. Marciniak RA, Lombard DB, Johnson FB et al. Nucleolar localization of the Werner syndrome protein in human cells. Proc Natl Acad Sci USA 1998; 95:6887-6892.
36. Ogburn CE, Oshima J, Poot M et al. An apoptosis-inducing genotoxin differentiates heterozygotic carriers for Werner helicase mutations from wild-type and homozygous mutants. Hum Genet 1997; 101:121-125.
37. Suzuki T, Shiratori M, Furuichi Y et al. Diverged nuclear localization of Werner helicase in human and mouse cells. Oncogene 2001; 20:2551-2558.
38. Shiratori M, Sakamoto S, Suzuki N et al. Detection by epitope-defined monoclonal antibodies of Werner DNA helicases in the nucleoplasm and their upregulation by cell transformation and immortalization. J Cell Biol 1999; 144:1-9.
39. Cox LS, Laskey RA. DNA replication occurs at discrete sites in pseudonuclei assembled from purified DNA in vitro. Cell 1991; 66:271-275.
40. Yan H, Newport J. An analysis of the regulation of DNA synthesis by cdk2, Cip1, and licensing factor. J Cell Biol 1995; 129:1-15.
41. Wold MS. Replication protein A: A heterotrimeric, single-stranded DNA-binding protein required for eukaryotic DNA metabolism. Annu Rev Biochem 1997; 66:61-92.

42. Brosh Jr RM, Orren DK, Nehlin JO et al. Functional and physical interaction between WRN helicase and human replication protein A. J Biol Chem 1999; 274:18341-18350.
43. Szekely AM, Chen YH, Zhang C et al. Werner protein recruits DNA polymerase delta to the nucleolus. Proc Natl Acad Sci USA 2000; 97:11365-11370.
44. Hanaoka F, Yamada M, Takeuchi F et al. Autoradiographic studies of DNA replication in Werner's syndrome cells. Adv Exp Med Biol 1985; 190:439-457.
45. Fujiwara Y, Kano Y, Ichihashi M et al. Abnormal fibroblast aging and DNA replication in the Werner syndrome. Adv Exp Med Biol 1985; 190:459-477.
46. Poot M, Hoehn H, Runger TM et al. Impaired S-phase transit of Werner syndrome cells expressed in lymphoblastoid cell lines. Exp Cell Res 1992; 202:267-273.
47. Matsumoto T, Shimamoto A, Goto M et al. Impaired nuclear localization of defective DNA helicases in Werner's syndrome. Nat Genet 1997; 16:335-336.
48. Matsumoto T, Imamura O, Goto M et al. Characterization of the nuclear localization signal in the DNA helicase involved in Werner's syndrome. Int J Mol Med 1998; 1:71-76.
49. Goto M, Miller RW, Ishikawa Y et al. Excess of rare cancers in Werner syndrome (adult progeria). Cancer Epidemiol Biomarkers Prev 1996; 5:239-246.
50. Monnat Jr RJ. Werner syndrome: Molecular genetics and mechanistic hypotheses. Exp Gerontol 1992; 27:447-453.
51. Schonberg S, Niermeijer MF, Bootsma D et al. Werner's syndrome: Proliferation in vitro of clones of cells bearing chromosome translocations. Am J Hum Genet 1984; 36:387-397.
52. Scappaticci S, Forabosco A, Borroni G et al. Clonal structural chromosomal rearrangements in lymphocytes of four patients with Werner's syndrome. Ann Genet 1990; 33:5-8.
53. Moser MJ, Holley WR, Chatterjee A et al. The proofreading domain of Escherichia coli DNA polymerase I and other DNA and/or RNA exonuclease domains. Nucleic Acids Res 1997; 25:5110-5118.
54. Shen JC, Loeb LA. The Werner syndrome gene: The molecular basis of RecQ helicase-deficiency diseases. Trends Genet 2000; 16:213-220.
55. Wang L, Ogburn CE, Ware CB et al. Cellular Werner phenotypes in mice expressing a putative dominant-negative human WRN gene. Genetics 2000; 154:357-362.
56. Constantinou A, Tarsounas M, Karow JK et al. Werner's syndrome protein (WRN) migrates Holliday junctions and colocalizes with RPA upon replication arrest. EMBO Reports 2000; 1:80-84.
57. Elli R, Chessa L, Antonelli A et al. Effects of topoisomerase II inhibition in lymphoblasts from patients with progeroid and "chromosome instability" syndromes. Cancer Genet Cytogenet 1996; 87:112-116.
58. Okada M, Goto M, Furuichi Y et al. Differential effects of cytotoxic drugs on mortal and immortalized B-lymphoblastoid cell lines from normal and Werner's syndrome patients. Biol Pharm Bull 1998; 21:235-239.
59. Poot M, Gollahon KA, Rabinovitch PS. Werner syndrome lymphoblastoid cells are sensitive to camptothecin-induced apoptosis in S-phase. Hum Genet 1999; 104:10-14.
60. Kallio M, Lahdetie J. Fragmentation of centromeric DNA and prevention of homologous chromosome separation in male mouse meiosis in vivo by topoisomerase II inhibitor etoposide. Mutagenesis 1996; 11:435-443.
61. Wang JC. DNA topoisomerases. Ann Rev Biochem 1985; 54:665-697.
62. Maybaum J, Bainnson AN, Roethel WM et al. Effects of incorporation of 6-thioguanine into SV40 DNA. Mol Pharmacol 1987; 32:606-614.
63. Uribe-Luna S, Quintana-Hau JD, Maldonado-Rodriguez R et al. Mutagenic consequences of the incorporation of 6-thioguanine into DNA. Biochem Pharmacol 1997; 54:419-424.
64. Fukuchi K, Tanaka K, Nakura J et al. Elevated spontaneous mutation rate in SV40-transformed Werner syndrome fibroblast cell lines. Somat Cell Mol Genet 1985; 11:303-308.
65. Fukuchi K, Martin GM, Monnat Jr RJ. Mutator phenotype of Werner syndrome is characterized by extensive deletions. Proc Natl Acad Sci USA 1989; 86:5893-5897.
66. Fukuchi K, Tanaka K, Kumahara Y et al. Increased frequency of 6-thioguanine-resistant peripheral blood lymphocytes in Werner syndrome patients. Hum Genet 1990; 84:249-252.
67. Lebel M. Werner syndrome: Genetic and molecular basis of a premature aging disorder. Cell Mol Life Sci 2001; 58:857-867.

68. Nunoshiba T, Demple B. Potent intracellular oxidative stress exerted by the carcinogen 4-nitroquinoline-N-oxide. Cancer Res 1993; 53:3250-3252.
69. Gebhart E, Bauer R, Raub U et al. Spontaneous and induced chromosomal instability in Werner syndrome. Hum Genet 1988; 80:135-139.
70. Ryan AJS, Squires S, Strutt HL et al. Camptothecin cytotoxicity in mammalian cells is associated with the induction of persistent double strand breaks in the replicating DNA. Nucleic Acids Res 1991; 19:3295-3300.
71. Lebel M, Cardiff RD, Leder P. Tumorigenic effect of nonfunctional p53 or p21 in mice mutant in the Werner syndrome helicase. Cancer Res 2001; 61:1816-1819.
72. Donehower LA, Harvey M, Slagle BL et al. Mice deficient for p53 are developmentally normal but susceptible to spontaneous tumours. Nature 1992; 356:215-221.
73. Wu J, He J, Mountz JD. Effect of age and apoptosis on the mouse homologue of the huWRN gene. Mech Ageing Dev 1998; 103:27-44.
74. Weirich HG, Weirich-Schwaiger H, Kofler H et al. Werner syndrome: studies in an affected family reveal a cellular phenotype of unaffected siblings. Mech Ageing Dev 1996; 88:1-15.
75. Wang L, Evans AE, Ogburn CE et al. Werner helicase expression in human fetal and adult aortas. Exp Gerontol 1999; 34:935-941.
76. Dimitrow PP, Czarnecka D, Kawecka-Jaszcz K et al. The frequency and functional impact of hypertension overlapping on hypertrophic cardiomyopathy: comparison between older and younger patients. J Hum Hypertens 1998; 12:633-634.
77. Goto M, Tanimoto K, Horiuchi Y et al. Family analysis of Werner's syndrome: A survey of 42 Japanese families with a review of the literature. Clin Genet 1981; 19:8-15.
78. Miki T, Nakura J, Ye L et al. Molecular and epidemiological studies of Werner syndrome in the Japanese population. Mech Ageing Dev 1997; 98:255-265.
79. Ye L, Miki T, Nakura J et al. Association of a polymorphic variant of the Werner helicase gene with myocardial infarction in a Japanese population. Am J Med Genet 1997; 68:494-498.
80. Oshima J. The Werner syndrome protein: An update. Bioessays 2000; 22:894-901.
81. Spillare EA, Robles AI, Wang XW et al. p53-mediated apoptosis is attenuated in Werner syndrome cells. Genes Dev 1999; 13:1355-1356.
82. Hollstein M, Sidransky D, Vogelstein B et al. p53 mutations in human cancers. Science 1991; 253:49-53.
83. Jacks T, Remington L, Williams BO et al. Tumor spectrum analysis in p53-mutant mice. Curr Biol 1994; 4:1-7.
84. Hinds PW, Weinberg RA. Tumor suppressor genes. Curr Opin Genet Dev 1994; 4:135-141.
85. El-Deiry WS, Tokino T, Velculescu VE et al. WAF1, a potential mediator of p53 tumor suppression. Cell 1993; 75:817-825.
86. Harper JW, Adami GR, Wei N et al. The p21 Cdk-interacting protein Cip1 is a potent inhibitor of G1 cyclin-dependent kinases. Cell 1993; 75:805-816.
87. Waga S, Hannon GJ, Beach D et al. The p21 inhibitor of cyclin-dependent kinases controls DNA replication by interaction with PCNA. Nature 1994; 369:574-578.
88. Noda A, Ning Y, Venable SF et al. Cloning of senescent cell-derived inhibitors of DNA synthesis using an expression screen. Exp Cell Res 1994; 211:90-98.
89. Kuerbitz SJ, Plunkett BS, Walsh WV et al. Wild-type p53 is a cell cycle checkpoint determinant following irradiation. Proc Natl Acad Sci USA 1992; 89:7491-7495.
90. Kemp CJ, Wheldon T, Balmain A. p53-deficient mice are extremely susceptible to radiation-induced tumorigenesis. Nat Genet 1994; 8:66-69.
91. Deng C, Zhang P, Harper JW et al. Mice lacking p21CIP1/WAF1 undergo normal development, but are defective in G1 checkpoint control. Cell 1995; 82:675-684.
92. Ohtsubo M, Theodoras AM, Schumacher J et al. Human cyclin E, a nuclear protein essential for the G1-to-S phase transition. Mol Cell Biol 1995; 15:2612-2624.
93. Levine AJ. p53, the cellular gatekeeper for growth and division. Cell 1997; 88:323-331.
94. Solomon E, Borrow J, Goddard AD. Chromosome aberrations and cancer. Science 1991; 254:1153-1160.
95. Weinberg RA. Tumor suppressor genes. Science 1991; 254:1138-1146.

96. Riggs JE. Aging, genomic entropy and carcinogenesis: implications derived from longitudinal age-specific colon cancer mortality rate dynamics. Mech Ageing Dev 1993; 72:165-181.
97. Van Alewijk DC, Van der Weiden MM, Eussen BJ et al. Identification of a homozygous deletion at 8p12-21 in a human prostate cancer xenograft. Genes Chromosomes Cancer 1999; 24:119-126.
98. Schiestl RH, Aubrecht J, Khogali F et al. Carcinogens induce reversion of the mouse pink-eyed unstable mutation. Proc Natl Acad Sci USA 1997; 94:4576-4581.
99. Jalili T, Murthy GG, Schiestl RH. Cigarette smoke induces DNA deletions in the mouse embryo. Cancer Res 1998; 58:2633-2638.
100. Lebel M. Increased frequency of DNA deletions in pink-eyed unstable mice carrying a mutation in the Werner Syndrome gene homologue. Carcinogenesis 2002; 23:213-216.
101. Leder A, Kuo A, Cardiff RD et al. v-Ha-ras transgene abrogates the initiation step in mouse skin tumorigenesis: effects of phorbol esters and retinoic acid. Proc Natl Acad Sci USA 1990; 87:9178-9182.
102. Leder A, Lebel M, Zhou F et al. Genetic Interaction between the unstable v-Ha-RAS transgene (Tg.AC) and the murine Werner Syndrome gene: Transgene instability and tumorigenesis. Oncogene 2002; 21:6657-6668.
103. Thompson KL, Rosenzweig BA, Sistare FD. An evaluation of the hemizygous transgenic Tg.AC mouse for carcinogenicity testing of pharmaceuticals. II. A genotypic marker that predicts tumorigenic responsiveness. Toxicol Pathol 1998; 26:548-555.
104. Thompson KL, Rosenzweig BA, Honchel R et al. Loss of critical palindromic transgene promoter sequence in chemically induced Tg.AC mouse skin papillomas expressing transgene-derived mRNA. Mol Carcinog 2001; 32:176-186.
105. Honchel R, Rosenzweig BA, Thompson KL et al. Loss of palindromic symmetry in Tg.AC mice with a nonresponder phenotype. Mol Carcinog 2001; 30:99-110.
106. Sinclair DA, Guarente L. Extrachromosomal rDNA circles—a cause of aging in yeast. Cell 1997; 91:1033-1042.
107. Yamagata K, Kato J, Shimamoto A et al. Bloom's and Werner's syndrome genes suppress hyperrecombination in yeast sgs1 mutant: implication for genomic instability in human diseases. Proc Natl Acad Sci USA 1998; 95:8733-8738.
108. Gangloff S, Soustelle C, Fabre F. Homologous recombination is responsible for cell death in the absence of the Sgs1 and Srs2 helicases. Nat Genet 2000; 25:192-194.
109. Ouellette MM, McDaniel LD, Wright WE et al. The establishment of telomerase-immortalized cell lines representing human chromosome instability syndromes. Hum Mol Genet 2000; 9:403-411.
110. Wyllie FS, Jones CJ, Skinner JW et al. Telomerase prevents the accelerated cell ageing of Werner syndrome fibroblasts. Nat Genet 2000; 24:16-17.
111. Cohen H, Sinclair DA. Recombination-mediated lengthening of terminal telomeric repeats requires the Sgs1 DNA helicase. Proc Natl Acad Sci USA 2001; 98:3174-379.
112. Cooper MP, Machwe A, Orren DK et al. Ku complex interacts with and stimulates the Werner protein. Genes Dev 2000; 14:907-912.
113. Li B, Comai L. Functional interaction between Ku and the werner syndrome protein in DNA end processing. J Biol Chem 2000; 275:28349-28352.
114. Morham SG, Kluckman KD, Voulomanos N et al. Targeted disruption of the mouse topoisomerase I gene by camptothecin selection. Mol Cell Biol 1996; 16:6804-6809.
115. Vogel H, Lim DS, Karsenty G et al. Deletion of Ku86 causes early onset of senescence in mice. Proc Natl Acad Sci USA 1999; 96:10770-10775.

CHAPTER 8

Proposed Biological Functions for the Werner Syndrome Protein in DNA Metabolism

Patricia L. Opresko, Jeanine A. Harrigan, Wen-Hsing Cheng, Robert M. Brosh, Jr. and Vilhelm A. Bohr

Abstract

Werner syndrome is a premature aging disease that is characterized by genomic instability. The gene defective in Werner syndrome encodes a protein with helicase and exonuclease activities. This chapter focuses on the proposed roles of the Werner syndrome protein in various aspects of DNA metabolism that are important for maintaining the genome integrity. Cellular and biochemical evidences for the participation of the Werner protein in various pathways including DNA replication, repair and recombination, are discussed.

Introduction

The complex clinical and cellular phenotype of Werner syndrome (WS) indicate that the WRN protein (WRN) may be involved in various aspects of DNA metabolism important for maintaining the integrity of the genome. Cloning of the WS gene has permitted biochemical approaches to investigating the role of WRN, including characterization of the catalytic activities and substrate specificities of the WRN protein. Another approach has been to determine protein interactors with WRN, and to let this guide us to the direction of the molecular pathways in which WRN participates. In this chapter we shall describe some of the protein interactions of WRN, and will discuss putative roles for these interactions in maintaining genome integrity in a biological setting. In some cases a physical association between WRN and a protein partner also results in a functional interaction, whereby the interacting proteins modulate each others activities (Table 1). The following is a discussion of the cellular and biochemical evidences for roles of the WRN protein in a number of different DNA metabolic functions.

DNA Replication

A number of WS cellular phenotypes suggest a role of the WRN protein in DNA replication. WS cells exhibit a reduced replicative life span[1,2] an extended S phase,[3] and a reduced frequency of replication initiation sites.[4,5] The interaction of WRN protein with a number of human nuclear proteins implicated in replication are consistent with the notion that WRN may be directly involved in the initiation and/or elongation modes of replicative DNA synthesis. Among the WRN protein interactors, the human single-stranded DNA binding protein replication protein A (RPA) is required for chromosomal replication[6] and enables WRN to unwind long DNA duplexes up to nearly a thousand base pairs.[7] The extensive sequence homology of WRN with *Xenopus laevis* replication focus-forming activity 1 (FFA1) suggests a role of WRN in replication initiation since the assembly of RPA-containing replication foci in *Xenopus laevis* egg extracts requires FFA1.[8] Aside from initiation, WRN helicase may function

Molecular Mechanisms of Werner's Syndrome, edited by Michel Lebel. ©2004 Eurekah.com and Kluwer Academic / Plenum Publishers.

Table 1. WRN-interacting proteins

WRN Interacting Proteins	Physical Interaction	Functional Interaction	Biological Role
RPA	Yes	Stimulates WRN helicase activity	DNA replication, DNA repair, DNA recombination
FEN-1	Yes	WRN stimulates FEN-1 flap cleavage activity	DNA replication, DNA repair, NHEJ
Pol-δ	Yes	WRN recruits p50 subunit of pol δ to the nucleolus, stimulates incorporation of nucleotides by pol δ	DNA replication, DNA repair
Topo I	Yes	unknown	DNA replication
PCNA	Yes	unknown	DNA replication, DNA repair
Ku 70/80	Yes	Stimulates WRN exonuclease activity	DSB repair, telomere maintenance
DNA-PK	Yes	Ku complexed with DNA-PKcs phosphorylates WRN and modulates WRN activities	DSB repair, telomere maintenance

Listed above are proteins that interact physically and/or functionally with the WRN protein. Possible cellular functions for the interactions are also shown.

during replication elongation. RPA is required for elongation of DNA synthesis,[6] but its role is not well understood. Genomic instability in WS cells, characterized by extensive deletions and chromosomal rearrangements, may arise due to basic defects in some aspect of DNA replication that involves coordinate action of WRN and RPA.

Recently it was reported that WRN physically and functionally interacts with the structure-specific flap endonuclease 1 (FEN-1), a protein that is implicated in Okazaki fragment processing.[9] The profound genomic instability of WS cells would be consistent with a defect in replication during lagging strand synthesis. A role of WRN in Okazaki fragment processing is further supported by the ability of the helicase to specifically displace 5' flap structures.[10] The interactions of WRN with DNA polymerase δ (pol δ),[11,12] topoisomerase I (topo I),[13] and proliferating cell nuclear antigen (PCNA)[14] support the finding that WRN is a component of a multi-protein complex[13] known as the replisome that functions at the replication fork. The precise molecular function(s) of WRN in a replication complex remains to be established.

WRN may have a specialized role after replication arrest. Genetic studies in bacteria suggest that RecQ functions with the 5' to 3' exonuclease RecJ at replication forks blocked by UV-induced DNA damage.[15] It was hypothesized that RecQ and RecJ processing involves the selective degradation of the nascent lagging DNA strand. Current models propose that RecQ displaces the nascent lagging strand, making it susceptible to nucleolytic degradation by RecJ. This allows the formation of a triple stranded structure that can be maintained by RecA until the lesion is repaired.[15] This would prevent deleterious strand breakage or recombination to maintain genomic stability. Similar to RecQ and RecJ, WRN and FEN-1 may function together during replication restart. WRN may insure efficient removal of the nascent lagging strand by virtue of its abilities to displace the nascent Okazaki fragment and stimulate the flap endonuclease and/or 5' to 3' exonuclease activities of FEN-1. A role of WRN in the cellular response to DNA damage is suggested by the translocation of WRN to sites of replication/

repair when replication is blocked by hydroxyurea.[16] WRN may function with other proteins at stalled replication forks or other specific structures to confer DNA structural stability in a rather dynamic environment that prevails in vivo. Biochemical characterization of the DNA substrate specificity of WRN helicase suggests that WRN is able to translocate on the lagging strand of a synthetic replication fork to unwind DNA duplex ahead of the fork.[10] This function of WRN may be important to the maintenance of fork progression when replication undergoes pausing as in the case of nucleotide depletion or arrest at the sites of DNA damage. Precisely how WRN acts in vivo to maintain genome stability remains to be defined, but it is highly likely to involve its helicase function on a structural intermediate of an important DNA metabolic pathway such as replication.

DNA Repair

Base Excision Repair

Base excision repair (BER) is a particularly important pathway because it corrects some of the most common DNA lesions including alkylated, oxidized, or deaminated bases. In the sub pathway called short patch BER, the repair gap is only one nucleotide, whereas in long patch repair the gap is 2-8 nucleotides in size.[17] Genomic instability of WS cells is consistent with a defect in DNA repair, and some evidence exists for a role of the WRN protein in BER. For example, a WRN$^{-/-}$ mutant chicken cell line shows hypersensitivity to methyl methanesulfonate (MMS), an agent which produces DNA damage that is repaired via BER, compared to the isogenic wild type cell line.[18] The WRN exonuclease is blocked by some oxidative DNA base lesions including 8-oxoguanine and 8-oxoadenine.[19] The WRN exonuclease can remove a single mismatched terminal nucleotide from a recessed 3' end,[20,21] and is active at nicks and gaps[21] suggesting a possible function for the WRN protein in the repair of oxidative DNA damage. Although the WRN protein does not bind with high affinity to DNA lesions, it may act as a sensor of their presence in DNA. This may be an early step in damage recognition, and WRN may recruit DNA repair enzymes to the site of the lesion via protein-protein interactions.

The WRN protein has been shown to interact physically and/or functionally with several proteins involved in long patch BER including pol δ, PCNA, RPA, and FEN-1. WRN interacts physically with pol δ and recruits it to the nucleolus.[12] Furthermore, WRN has been shown to stimulate incorporation of nucleotides by pol δ.[11] WRN also interacts physically with PCNA,[13] however, a functional interaction has yet to be determined. RPA stimulates the WRN helicase to unwind long duplex DNA substrates that WRN cannot unwind alone.[7,22] Furthermore, the WRN protein interacts physically with FEN-1 and stimulates its flap cleavage activity.[9] FEN-1 participates in both DNA replication and BER. In light of the fact that RPA has been shown to stimulate the long patch pathway of BER [23,24] and FEN-1 is necessary for cleavage of the flap generated, it is interesting to speculate that WRN could participate with RPA to unwind BER DNA intermediates as well as stimulate FEN-1 flap cleavage.

Although repair of apurinic/apyrimidinic (AP) sites by whole cell extracts from WS cells appears to be normal,[25] the role of WRN in long patch BER remains to be determined. The substrates on which the WRN helicase and exonuclease act, as well as the proteins with which WRN interacts, strongly suggest that WRN may participate in BER.

Double Strand Break Repair

DNA double strand breaks (DSBs) are introduced into the genome during normal DNA metabolism or by exogenous sources including ionizing radiation and oxidative stress. Homologous recombination and non-homologous end joining (NHEJ) processes are necessary to prevent lethality induced by DSBs, and act to repair the breaks. WS cells are mildly sensitive to ionizing radiation, an agent that induces DSBs, compared to normal cells.[26] However, WS cells are hypersensitive to DNA cross-linking agents that ultimately result in DSBs.[27] In addition, WS cells are characterized by non-homologous chromosome exchanges, termed variegated translocation mosaicism, and large chromosomal deletions[28,29] consistent with possible defects in end-joining.

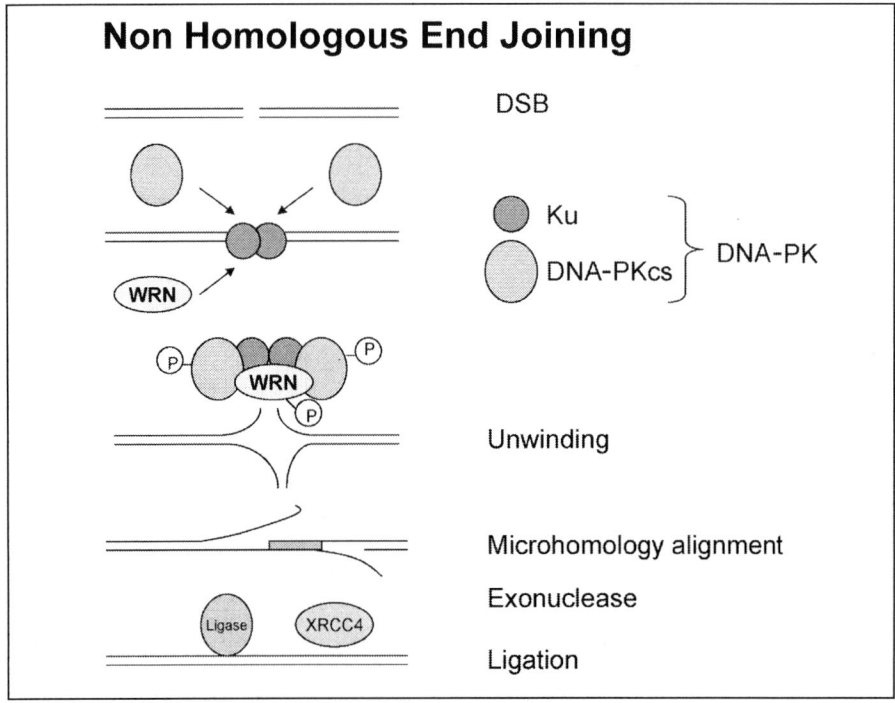

Figure 1. This model shows how WRN may act with the DNA-PK complex in NHEJ. Presumably the Ku heterodimer binds the broken DNA ends and then recruits other enzyme factors to the site of repair, including DNA-PKcs and WRN. Ku and DNA-PKcs form the DNA-PK kinase which autophosphorylates and in turn can phosphorylate WRN. A DNA unwinding step takes place at the broken ends and the ends anneal at sites of microhomology. Unannealed ends generate single stranded flaps that are presumably removed by an exonuclease. The WRN helicase and exononulease activities may be involved in both end processing steps. The DNA-PK complex can potentially modulate the relative WRN activities through interaction with and phosphorylation of the WRN protein. After the ends are processed and the gaps are filled in, the remaining nicks are sealed by a DNA ligase IV complexed with XRCC4.

Non-Homologous End Joining

The WRN protein interacts physically and functionally with several key components of the NHEJ pathway. A critical player in NHEJ is the DNA-dependent protein kinase complex (DNA-PK), which consists of the catalytic subunit of the DNA-PK (DNA-PKcs) and the Ku 70/80 heterodimer. Both components are required for repair of DSBs induced by reactive oxygen species generated by metabolic processes and ionizing radiation.[30] Current models suggest that Ku binds to the broken DNA double strand ends and then recruits DNA-PKcs to form the active protein kinase DNA-PK (Fig. 1). A DNA unwinding step takes place and the ends anneal at sites of microhomology. In this process, the single stranded flaps that are generated may be removed by an exonuclease. Thus, both a DNA helicase and an exonuclease activity are involved, and the WRN protein possesses both of these catalytic functions. Furthermore, Ku physically interacts with the WRN protein and stimulates the WRN 3' to 5' exonuclease activities on both double stranded and single stranded DNA molecules.[31,32] In addition, two independent studies have found that WRN forms a complex with DNA-PK, and that DNA-PK phosphorylates the WRN protein both in vitro and in vivo.[26,33] These studies indicate that the interaction with and phosphorylation of WRN by the DNA-PK complex modulates the WRN helicase and exonuclease activities. Evidence in yeast also indicates that another WRN protein partner, FEN-1, may function in NHEJ to process DNA ends prior to ligation.[34] Furthermore, in a cellular assay that measures NHEJ with a linear plasmid, WS cells display extensive deletions at the nonhomologous joined ends, compared to normal cells.[35] Clearly, more work is required to determine whether defective end-joining contributes to the genomic instability that is characteristic of WS.

DNA Recombination and Recombinational Repair

A putative role of the human WRN protein in recombination repair was suggested by studies using WRN homologs in bacteria and yeast. The *E. coli* WRN homolog RecQ can suppress illegitimate recombination through initiating homologous recombination and disrupting joint molecules.[36] Similar functions of the *Saccharomyces cerevisiae* RecQ homolog Sgs1 in recombinational repair have also been described.[37] In yeast *sgs1* mutants, the elevated rate of homologous recombination can be suppressed by overexpressing WRN.[38] Recent studies using WS cells support a role for human WRN in recombinational repair. Cells from WS patients show increased rates of mitotic recombination.[39] Results from this study provide evidence that the resolution step of recombination intermediates is defective in WS cells, and not the initiation step of homologous recombination. Furthermore, WS cells are sensitive to agents that induce interstrand crosslinks,[25,40] and these defects are repairable by homologous recombination. A model for the function of WRN in recombinational repair has been proposed[41] whereby WRN functions in association with Rad51, a single strand binding protein, at a stalled replication fork caused by DNA damage in one DNA strand. WRN may act in concert with Rad51 and/or RPA, another single strand binding protein, to facilitate strand exchange and Holiday junction formation and/or migration. After resolution of the recombination intermediate, the lesion now resides in duplex DNA and is removed by repair enzymes. Consistent with this model, WRN relocalizes into nuclear foci and co-localizes with RPA and Rad51 upon treatment with various DNA damaging agents that cause DSBs.[42] In addition, WRN interacts physically and functionally with RPA[7] and is able to dissociate the Holliday junction recombination intermediate.[16] Furthermore, evidence in vitro indicates that the WRN helicase and exonuclease may act coordinately to remove repair and recombination intermediates.[43] Although these studies provide strong evidence in support of a role for human WRN in recombinational repair, the precise molecular mechanism by which WRN participates in this process awaits further investigation.

Transcription

In addition to defects in DNA repair and replication, defects in transcription have also been detected in some WS cell lines. WS lymphoblasts containing homozygous mutations in the *WRN* gene display a 40-60% reduction in transcription efficiency compared to normal cells.[44] This reduction is observed both in vivo using a [^{3}H]-uridine incorporation assay with permeabilized cells, and in vitro with nuclear extracts and a plasmid template bearing an RNA polymerase II-specific promoter. The observed transcription deficiency can be complemented by the addition of normal cell extract to the chromatin from WS cells. Furthermore, the addition of wild type WRN protein stimulates RNA pol II-dependent transcription in the in vitro assay, an affect not observed with mutant WRN proteins that either lack the 27-amino acid direct repeat or contain a point mutation in the helicase domain (K577M). A role for WRN in transcription is also suggested by the observation that overexpression of WRN results in enhanced p53-dependent transcriptional activity.[45] These observations indicate that the WRN protein may act as a general activator of RNA polymerase II transcription.

There is also some evidence that WRN may participate in RNA pol I-dependent transcription of ribosomal RNA. Several reports have indicated that WRN localizes to trancriptionally active nucleoli.[46-48] WRN migrates out of the nucleoli when rRNA transcription is inhibited by treatment with actinomycin D.[49] Furthermore, WRN coimmunoprecipitates with a subunit of the RNA pol I complex.[49] This same study reported that WS fibroblasts display decreased levels of rRNA, which can be rescued by expression of exogenous WRN. However, a previous study did not observe any difference in the steady-state levels of 28S rRNA during the life span of WS fibroblasts, compared to normal cells, although methylation of rRNA genes was accelerated in the WS cells.[50] Further investigation is required to define the precise roles of WRN in RNA pol I and pol II mediated transcription.

Telomere Metabolism

Defects in the WS protein are associated with genetic instability and a decline in proliferative competence. Similarly, consequences of telomere dysfunction include replicative senescence and/or genomic instability.[51] Telomeres cap and protect chromosome ends and the progressive decline in telomere lengths that occurs with each cell division eventually triggers replicative senescence.[52] However, telomere-associated senescence can be bypassed by the action of telomerase which extends telomeres.[53] The forced expression of exogenous telomerase in WS fibroblasts was found to extend the cellular life span, thus suggesting that the premature senescent phenotype in WS cells may be related to telomere dysfunction.[54,55] Furthermore, the expression of exogenous telomerase also partially reversed the hypersensitivity to 4-nitroquinoline-1-oxide exhibited by WS cell lines.[56] The apparent connection between the WRN protein and telomerase is unknown.

The WRN protein may participate in DNA metabolic processes at telomeric ends, including replication, repair and/or recombination that affect telomere integrity. The decline in proliferative capacity of WS cells cannot be explained simply by acceleration of telomeric loss. Although WS fibroblasts display accelerated rates of telomere shortening, at senescence the mean telomere length is longer in WS cells compared to senescent controls.[57] However, studies in which telomere lengths were altered by manipulation of telomere maintenance proteins TRF2 and TIN2, indicate that changes in telomere structure, rather than telomere length, trigger replicative senescence.[58,59] Electron microscopy studies have shown that telomeres in mammalian cells form t loop structures in which the single-stranded 3' tail invades the homologous duplex telomere region and creates a displacement loop (D-loop).[60] Consequently, the telomeric end is protected and sequestered. Presumably, structures at telomeric ends must be resolved in order for telomerase, and for DNA replication and repair proteins, to gain access to the terminal region of the telomere. The WRN protein is a likely candidate to participate in this process, since it unwinds various DNA secondary structures and recombination intermediates[16,61,62] and is predicted to resolve replication fork blocks (see the DNA REPLICATION section).

Consistent with a role for WRN in telomere maintenance, WS fibroblasts were observed to be defective in repair of UV lesions at telomeres, compared to wild type.[57] As mentioned earlier, WRN interacts with the Ku heterodimer and is phosphorylated by DNA-PK (see section on NHEJ). Ku and DNA-PKcs have been found to localize to telomeres and defects in these proteins lead to dysfunctional telomeres and accelerated telomere shortening.[63,64] Furthermore, Ku physically interacts with the important telomere maintenance proteins TRF2 and TRF1.[65,66] How Ku and DNA-PKcs act in telomere maintenance is not yet known, but they likely function in the cellular response to dysfunctional and/or damaged telomeres, perhaps in a pathway with WRN.

Further support for a role of WRN in telomere metabolism is derived from studies in yeast. Three independent studies indicate that the RecQ homolog in *Saccharomyces cerevisiae*, Sgs1, participates in a telomerase-independent pathway for telomere lengthening.[67-69] The mechanism of this pathway is not well understood, but is predicted to involve recombination.[70] Evidence for a similar pathway has been found in telomerase-negative immortalized mammalian cells.[71] In these cell lines WRN co-localizes with telomeric DNA and the telomere TRF1 and TRF2 proteins,[67] suggesting WRN may participate in recombination based pathways at telomeric ends.

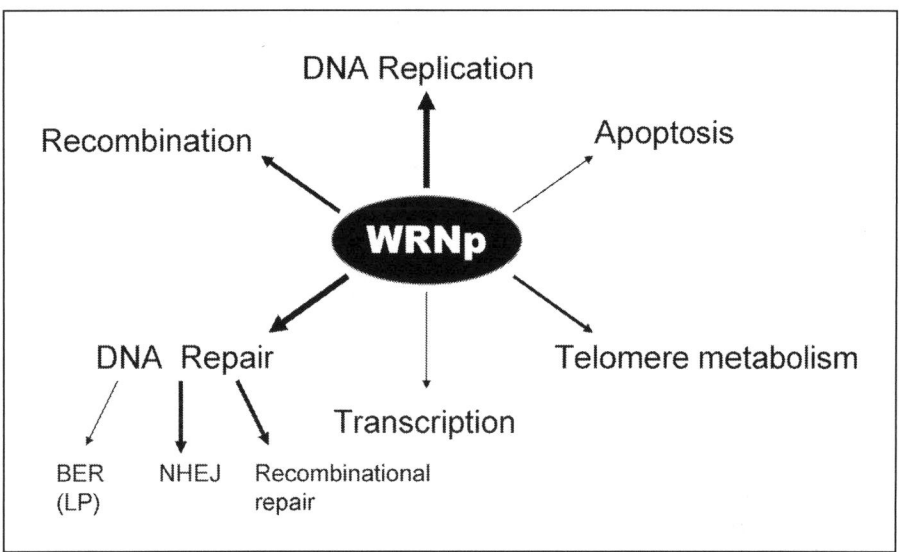

Figure 2. A summary of the DNA metabolic pathways WRN may be involved in are shown. Some indication of the hypothesized relative importance of WRN in these pathways is indicated by the density of the arrows used. The phenotypes of the WS cell lines and the biochemical analysis the WRN protein indicate that WRN likely functions in multiple pathways that are important for maintaining the genome integrity. The precise role of WRN in each of these pathways has yet to be determined. (LP)= long patch.

Summary and Perspectives

The complex clinical and cellular phenotypes of WS indicate that the WRN protein is critically important for maintaining the genome and has a profound influence on cellular life span. Significant progress has been made toward understanding the biochemical characteristics of the Werner syndrome protein, including its catalytic activities, substrate specificities and protein interactors. Taken together, the phenotypes of the WS cell lines and the biochemical analysis of the WRN protein indicate that WRN likely functions in multiple DNA related pathways. A summary of the pathways WRN may be involved in is shown in Figure 2. Some indication of WRN's relative importance in these pathways is indicated by the density of the arrows used. Based on Figure 2 and on the above discussion, it is evident that the major role of WRN lies in maintaining genomic stability, and that further investigation is required to determine the relative contribution and importance of WRN function in DNA replication, repair and recombination pathways. An understanding of precisely how WRN functions in vivo is necessary to better understand how defects in WRN contribute to the accelerated aging phenotypes and genomic instability observed in WS patients. Given that so many of the clinical features of WS are early symptoms of normal aging, it is likely that lessons learned from Werner syndrome should yield insights into understanding the normal aging process.

References

1. Salk D, Bryant E, Hoehn H et al. Growth characteristics of Werner syndrome cells in vitro. Adv Exp Med Biol 1985; 190:305-311.
2. Martin GM, Sprague CA, Epstein CJ. Replicative life-span of cultivated human cells. Effects of donor's age, tissue, and genotype. Lab Invest 1970; 23:86-92.
3. Poot M, Hoehn H, Runger TM et al. Impaired S-phase transit of Werner syndrome cells expressed in lymphoblastoid cells. Exp Cell Res 1992; 202:267-73.
4. Takeuchi F, Hanaoka F, Goto M et al. Altered frequency of initiation sites of DNA replication in Werner's syndrome cells. Hum Genet 1982; 60:365-368.

5. Hanaoka F, Yamada M, Takeuchi F et al. Autoradiographic studies of DNA replication in Werner's syndrome cells. Adv Exp Med Biol 1985; 190:439-457.
6. Wold MS. Replication protein A: A heterotrimeric, single-stranded DNA-binding protein required for eukaryotic DNA metabolism. Annu Rev Biochem 1997; 66:61-92.
7. Brosh Jr RM, Orren DK, Nehlin JO et al. Functional and physical interaction between WRN helicase and human replication protein A. J Biol Chem 1999; 274:18341-18350.
8. Yan H, Chen C-Y, Kobayashi R et al. Replication focus-forming activity 1 and the Werner syndrome gene product. Nat Genet 1998; 19:375-378.
9. Brosh Jr RM, von Kobbe C, Sommers JA et al. Werner syndrome protein interacts with human flap endonuclease 1 and stimulates its cleavage activity. EMBO J 2001; 20:5791-5801.
10. Brosh Jr RM, Waheed J, Sommers JA. Biochemical characterization of the DNA substrate specificity of werner syndrome helicase. J Biol Chem 2002; 277:23236-23245.
11. Kamath-Loeb AS, Johansson E, Burgers PM et al. Functional interaction between the Werner Syndrome protein and DNA polymerase delta. Proc Natl Acad Sci USA 2000; 97:4603-4608.
12. Szekely AM, Chen YH, Zhang C et al. Werner protein recruits DNA polymerase delta to the nucleolus. Proc Natl Acad Sci USA 2000; 97:11365-11370.
13. Lebel M, Spillare EA, Harris CC et al. The werner syndrome gene product Co-purifies with the DNA replication complex and interacts with PCNA and topoisomerase I. J Biol Chem 1999; 274:37795-37799.
14. Lebel M, Leder P. A deletion within the murine Werner syndrome helicase induces sensitivity to inhibitors of topoisomerase and loss of cellular proliferative capacity. Proc Natl Acad Sci USA 1998; 95:13097-13102.
15. Courcelle J, Hanawalt PC. RecQ and RecJ process blocked replication forks prior to the resumption of replication in UV-irradiated Escherichia coli. Mol Gen Genet 1999; 262:543-551.
16. Constantinou A, Tarsounas M, Karow JK et al. Werner's syndrome protein (WRN) migrates Holliday junctions and co- localizes with RPA upon replication arrest. EMBO Rep 2000; 1:80-84.
17. Krokan HE, Nilsen H, Skorpen F et al. Base excision repair of DNA in mammalian cells. FEBS Lett 2000; 476:73-77.
18. Imamura O, Fujita K, Itoh C et al. Werner and Bloom helicases are involved in DNA repair in a complementary fashion. Oncogene 2002; 21:954-963.
19. Machwe A, Ganunis R, Bohr VA et al. Selective blockage of the 3'→5' exonuclease activity of WRN protein by certain oxidative modifications and bulky lesions in DNA. Nucleic Acids Res 2000; 28:2762-2770.
20. Kamath-Loeb AS, Shen JC, Loeb LA et al. Werner Syndrome Protein. Ii. characterization of the integral 3' → 5' dna exonuclease. J Biol Chem 1998; 273:34145-34150.
21. Huang S, Beresten S, Li B et al. Characterization of the human and mouse WRN 3'→5' exonuclease. Nucleic Acids Res 2000; 28:2396-2405.
22. Shen JC, Gray MD, Oshima J et al. Characterization of Werner syndrome protein DNA helicase activity: directionality, substrate dependence and stimulation by replication protein A. Nucleic Acids Research 1998; 26:2879-2885.
23. Dianov GL, Jensen BR, Kenny MK et al. Replication Protein A Stimulates Proliferating Cell Nuclear Antigen- Dependent Repair of Abasic Sites in DNA by Human Cell Extracts. Biochemistry 1999; 38:11021-11025.
24. DeMott MS, Zigman S, Bambara RA. Replication protein A stimulates long patch DNA base excision repair. J Biol Chem 1998; 273:27492-27498.
25. Bohr VA, Souza PN, Nyaga SG et al. DNA repair and mutagenesis in Werner syndrome. Environ Mol Mutagen 2001; 38:227-234.
26. Yannone SM, Roy S, Chan DW et al. Werner syndrome protein is regulated and phosphorylated by dna- dependent protein kinase. J Biol Chem 2001; 276:38242-38248.
27. Poot M, Yom JS, Whang SH et al. Werner syndrome cells are sensitive to DNA cross-linking drugs. FASEB J 2001; 15:1224-1226.
28. Fukuchi K, Martin GM, Monnat RJ. Mutator phenotype of Werner syndrome is characterized by extensive deletions. Proc Nat Acad Sci USA 1989; 86:5893-5897.
29. Stefanini M, Scappaticci S, Lagomarsini P et al. Chromosome instability in lymphocytes from a patient with Werner's syndrome is not associated with DNA repair defects. Mutation Res 1989; 219:179-185.
30. Featherstone C, Jackson SP. Ku, a DNA repair protein with multiple cellular functions? Mutat Res 1999; 434:3-15.
31. Cooper MP, Machwe A, Orren DK et al. Ku complex interacts with and stimulates the Werner protein. Genes Dev 2000; 14:907-912.

32. Li B, Comai L. Requirements for the nucleoytic processing of DNA ends by the Werner syndrome protein:Ku70/80 complex. J Biol Chem 2001; 276:9896-9902.
33. Karmakar P, Piotrowski J, Brosh RM Jr et al. Werner protein is a target of DNA-dependent protein kinase in vivo and in vitro, and its catalytic activities are regulated by phosphorylation. J Biol Chem 2002; 277:18291-18302.
34. Wu X, Wilson TE, Lieber MR. A role for FEN-1 in nonhomologous DNA end joining: the order of strand annealing and nucleolytic processing events. Proc Natl Acad Sci USA 1999; 96:1303-1308.
35. Oshima J, Huang S, Pae C et al. Lack of WRN results in extensive deletion at nonhomologous joining ends. Cancer Res 2002; 62:547-551.
36. Harmon FG, Kowalczykowski SC. RecQ helicase, in concert with RecA and SSB proteins, initiates and disrupts DNA recombination. Genes Dev 1998; 12:1134-1144.
37. Myung K, Datta A, Chen C et al. SGS1, the Saccharomyces cerevisiae homologue of BLM and WRN, suppresses genome instability and homeologous recombination. Nat Genet 2001; 27:113-116.
38. Yamagata K, Kato J, Shimamoto A et al. Bloom's and Werner's syndrome genes suppress hyperrecombination in yeast sgs1 mutant: Implication for genomic instability in human diseases. Proc Natl Acad Sci USA 1998; 95:8733-8738.
39. Prince PR, Emond MJ, Monnat RJ Jr. Loss of Werner syndrome protein function promotes aberrant mitotic recombination. Genes Dev 2001; 15:933-938.
40. Poot M, Gollahon KA, Emond MJ et al. Werner syndrome diploid fibroblasts are sensitive to 4-nitroquinoline-N- oxide and 8-methoxypsoralen: implications for the disease phenotype. FASEB J 2002; 16:757-758.
41. Shen JC, Loeb LA. The Werner syndrome gene: The molecular basis of RecQ helicase-deficiency diseases. Trends Genet 2000; 16:213-220.
42. Sakamoto S, Nishikawa K, Heo SJ et al. Werner helicase relocates into nuclear foci in response to DNA damaging agents and co-localizes with RPA and Rad51. Genes Cells 2001; 6:421-430.
43. Opresko PL, Laine JP, Brosh Jr RM et al. Coordinate Action of the Helicase and 3' to 5' Exonuclease of Werner Syndrome Protein. J Biol Chem 2001; 276:44677-44687.
44. Balajee AS, Machwe A, May A et al. The Werner syndrome protein is involved in RNA polymerase II transcription. Mol Biol Cell 1999; 10:2655-2668.
45. Blander G, Kipnis J, Leal JF et al. Physical and functional interaction between p53 and the Werner's syndrome protein. J Biol Chem 1999; 274:29463-29469.
46. Gray MD, Wang L, Youssoufian H et al. Werner helicase is localized to transcriptionally active nucleoli of cycling cells. Exp Cell Res 1998; 242:487-494.
47. Marciniak RA, Lombard DB, Johnson FB et al. Nucleolar localization of the Werner syndrome protein in human cells. Proc Natl Acad Sci USA 1998; 95:6887-6892.
48. Suzuki T, Shiratori M, Furuichi Y et al. Diverged nuclear localization of Werner helicase in human and mouse cells. Oncogene 2001; 20:2551-2558.
49. Shiratori M, Suzuki T, Itoh C et al. WRN helicase accelerates the transcription of ribosomal RNA as a component of an RNA polymerase I-associated complex. Oncogene 2002; 21:2447-2454.
50. Machwe A, Orren DK, Bohr VA. Accelerated methylation of ribosomal RNA genes during the cellular senescence of werner syndrome fibroblasts. FASEB J 2000; 14:1715-1724.
51. Campisi J, Kim S, Lim CS et al. Cellular senescence, cancer and aging: the telomere connection. Exp Gerontol 2001; 36:1619-1637.
52. Allsopp RC, Vaziri H, Patterson C et al. Telomere length predicts replicative capacity of human fibroblasts. Proc Natl Acad Sci USA 1992; 89:10114-10118.
53. Bodnar AG, Ouellette M, Frolkis M et al. Extension of life-span by introduction of telomerase into normal human cells. Science 1998; 279:349-352.
54. Wyllie FS, Jones CJ, Skinner JW et al. Telomerase prevents the accelerated cell ageing of Werner syndrome fibroblasts. Nat Genet 2000; 24:16-17.
55. Ouellette MM, McDaniel LD, Wright WE et al. The establishment of telomerase-immortalized cell lines representing human chromosome instability syndromes. Hum Mol Genet 2000; 9:403-411.
56. Hisama FM, Chen YH, Meyn MS et al. WRN or telomerase constructs reverse 4-nitroquinoline 1-oxide sensitivity in transformed Werner syndrome fibroblasts. Cancer Res 2000; 60:2372-2376.
57. Kruk PA, Rampino NJ, Bohr VA. DNA damage and repair in telomeres: relation to aging. Proc Natl Acad Sci U S A 1995; 92:258-262.
58. Karlseder J, Smogorzewska A, De Lange T. Senescence induced by altered telomere state, not telomere loss. Science 2002; 295:2446-2449.
59. Rubio MA, Kim SH, Campisi J. Reversible manipulation of telomerase expression and telomere length: Implications for the ionizing radiation response and replicative senescence of human cells. J Biol Chem 2002; 277(32):28609-17.

60. Griffith JD, Comeau L, Rosenfield S et al. Mammalian telomeres end in a large duplex loop. Cell 1999; 97:503-514.
61. Brosh Jr RM, Majumdar A, Desai S et al. Unwinding of a DNA triple helix by the werner and bloom syndrome helicases. J Biol Chem 2001; 276:3024-3030.
62. Mohaghegh P, Karow JK, Brosh Jr RM et al. The Bloom's and Werner's syndrome proteins are DNA structure-specific helicases. Nucleic Acids Res 2001; 29:2843-2849.
63. Bailey SM, Meyne J, Chen DJ et al. DNA double-strand break repair proteins are required to cap the ends of mammalian chromosomes. Proc Natl Acad Sci USA 1999; 96:14899-14904.
64. d'Adda dF, Hande MP, Tong W et al. Effects of DNA nonhomologous end-joining factors on telomere length and chromosomal stability in mammalian cells. Curr Biol 2001; 11:1192-1196.
65. Song K, Jung D, Jung Y et al. Interaction of human Ku70 with TRF2. FEBS Lett 2000; 481:81-85.
66. Hsu HL, Gilley D, Galande SA et al. Ku acts in a unique way at the mammalian telomere to prevent end joining. Genes Dev 2000; 14:2807-2812.
67. Johnson FB, Marciniak RA, McVey M et al. The Saccharomyces cerevisiae WRN homolog Sgs1p participates in telomere maintenance in cells lacking telomerase. EMBO J 2001; 20:905-913.
68. Cohen H, Sinclair DA. Recombination-mediated lengthening of terminal telomeric repeats requires the Sgs1 DNA helicase. Proc Natl Acad Sci USA 2001; 98:3174-3179.
69. Huang P, Pryde FE, Lester D et al. SGS1 is required for telomere elongation in the absence of telomerase. Curr Biol 2001; 11:125-129.
70. Henson JD, Neumann AA, Yeager TR et al. Alternative lengthening of telomeres in mammalian cells. Oncogene 2002; 21:598-610.
71. Bryan TM, Englezou A, Dalla-Pozza L et al. Evidence for an alternative mechanism for maintaining telomere length in human tumors and tumor-derived cell lines. Nat Med 1997; 3:1271-1274.

CHAPTER 9

Replicative Senescence, Telomeres and Werner's Syndrome

Richard G.A. Faragher

Abstract

Werner's syndrome (WS) is studied as a model of accelerated aging and results from mutations in a *recQ* helicase (*WRN*). WS fibroblasts show a mutator phenotype (producing large DNA deletions), replication fork stalling, increased rates of mean telomeric loss and accelerated replicative senescence. If Werner's syndrome is to be of value as a model disease for the study of human aging, it is necessary to determine which of these features is relevant to the normal process and in what way. This requires that the biology of the disease be placed in the wider context of mechanisms believed to contribute to the normal aging process. The replicative senescence of somatic cells has been proposed as one such candidate mechanism for the aging of mitotic tissue. However, not all mitotic tissues are affected in Werner's syndrome. Does this apparent paradox indicate that Werner's syndrome is a poor model for aging or that replicative senescence is an unsatisfactory mechanism for it in many tissues? These questions are discussed with reference to the manner in which senescent cells are believed to contribute to the aging process and the ways in which cells may count divisions. Perhaps unexpectedly, the available data suggest that accelerated senescence should not be observed in all mitotic WS cell types.

Introduction

Most of the chapters in this book are devoted to the molecular, biochemical and genetic detail of Werner's syndrome, the *WRN* gene product and RecQ helicases in general. This chapter will not attempt to add further detail to these studies but instead will attempt to complement them by setting the biology of Werner's syndrome (considered in its broadest terms) in the context of the aging of normal organisms. In 1980, and even in 1990, such an endeavor would probably have been rather unnecessary. At that time, the level of data concerning the causal molecular processes operating in Werner's syndrome was relatively slight and the view of workers in the field was almost required to be correspondingly broad. Today, however, thanks in no small part to the other contributors to this volume, this is no longer the case. It is now possible for a worker entering the Werner's syndrome field for the first time, to study the biology of the WRN helicase almost in isolation and certainly without any detailed clinical knowledge of the disease itself. Much is being gained as a result of this ability to specialize. However, it is also incumbent on us to see that as little as possible of the "big picture" is lost.

Why Do Gerontologists Study Werner's Syndrome?

As an evolutionary side effect of millions of years of selection for reproductive success, the genetic basis of aging is potentially extremely broad. Perhaps as much as 7% of the total genome may play a role in determining the lifespan of humans.[1] With such a highly polygenic

Molecular Mechanisms of Werner's Syndrome, edited by Michel Lebel. ©2004 Eurekah.com and Kluwer Academic / Plenum Publishers.

system, to suggest that any single mechanism or simple network of mechanisms is responsible for the aging of a whole organism is rather naive. By the same logic the identification of candidate genes involved in successful aging through the study of normal centenarian "survivor" populations, whilst potentially feasible, is likely to prove an extremely complicated task. An alternative approach, first articulated as a formal concept by George Martin, is the study of heritable genetic diseases which mimic some, but not all, of the features of the aging process in order to gain insights into how the aging process functions in normal individuals. The study of such "progeroid" syndromes, of which Werner's syndrome is the classic example, has the advantage that only a single gene is usually involved in each case. This renders hypotheses easier to frame and test and makes the manipulation of the candidate allele possible in a way denied to workers studying highly polygenic traits. The disadvantage of studying progeroid syndromes is that they are essentially phenocopies of normal aging rather than the genuine article. Any observations made using them must thus be evaluated within the context of theories designed to explain how normal aging operates. It is my view that the data available to us from the study of Werner's syndrome have been extremely valuable in advancing our understanding of the normal aging process in two distinct ways. Firstly, consideration of the disease has been useful in a methodological sense because it has driven the formulation of extremely precise hypotheses that are amenable to clear refutation and are thus (in Karl Popper's sense of the term) extremely powerful.[2] Secondly, data arising from the study of Werner's syndrome have been practically informative in distinguishing between (or refining) hypotheses already widespread within gerontology. In my view, the existing data from Werner's syndrome:

　i. Are consistent with the hypothesis that the replicative senescence of human cells plays a causal role in normal human aging.
　ii. Provide visible limits to the replicative senescence hypothesis that emphasize the multicausal nature of the aging process.
　iii. Stress the context dependency of senescence mechanisms both between different tissues in the same organism and between the same tissue in different species. This latter element has implications not simply outside the study of the disease, but outside gerontology as well.

The Replicative Senescence Hypothesis of Aging

Having asserted that the available data from Werner's syndrome are most consistent with the hypothesis that replicative (or cell) senescence plays a role in human aging, justification of that statement requires an explanation of the cell senescence hypothesis in some detail. A priori, the simplest question which can be asked regarding the operation of the aging process is whether the mechanism of aging functions at the level of the organism or of the cells which compose it. For the first half of this century, it was generally believed (based upon experiments conducted in the laboratory of the Nobel laureate Alexis Carrel) that individual cells were immortal once placed into culture and that the mechanism of the aging process essentially represented a failure of the organization rather than the component parts.[3] However, in the 1960s, a series of experiments by Hayflick and Moorhead demonstrated that normal human fibroblasts would only proliferate for a fixed number of passages during which the population would double in number about 50 times.[4,5] The number of "population doublings" (PD) which a culture would undergo appeared to be fixed by some mechanism internal to the cells which composed it. This concept of an internal system for "counting" cell divisions was based upon experiments in which cells from populations with different growth abilities were cocultivated. It was also based on the observation that cryopreservation in liquid nitrogen did not "reset" the culture and did not allow it to proliferate indefinitely.[6] After completing its quota of population doublings, the culture would be entirely composed of cells in a non-dividing state which Hayflick termed "senescence". Based in part upon an initial observation that embryonic fibroblasts completed more PDs than neonatal cells, Hayflick proposed that this replicative "senescence" was linked

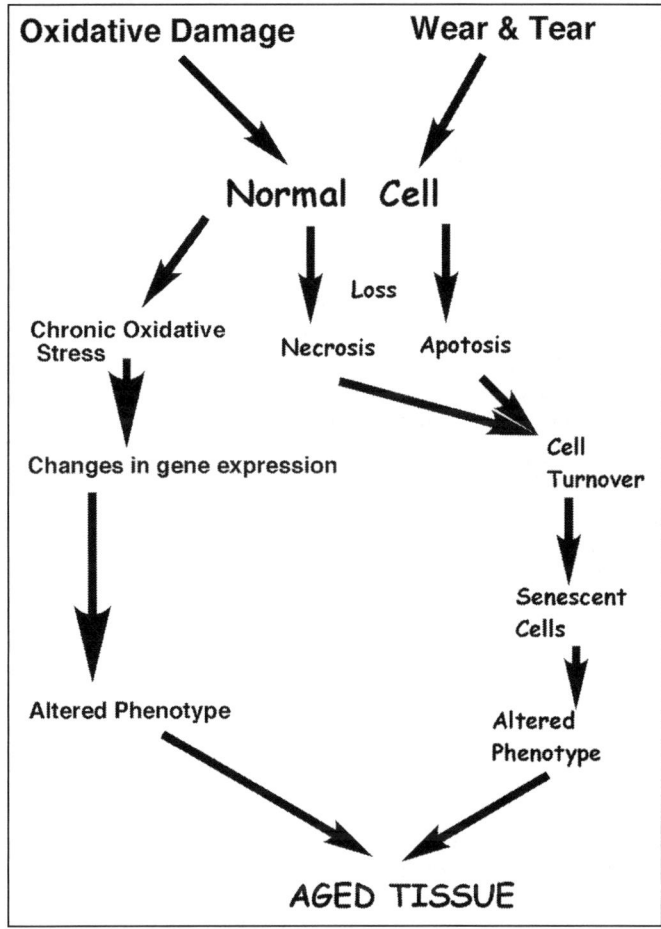

Figure 1. Simple schematic comparing the cell senescence hypothesis of aging (right hand flow) with the dysdifferentiation hypothesis of aging (left hand flow). The cell senescence hypothesis postulates that in the normal course of life there is cell loss. That loss is balanced by cell division which is actively monitored. One or more "replicometers" act to trigger permanent cell cycle exit (senescence) in individual cells (see text). Cell cycle exit is associated with a broad alteration in gene expression leading to an altered phenotype that affects the microenvironment in which the cell resides and ultimately the entire tissue. In the dysdifferentiation model, chronic oxidative stress leads to a regearing of gene expression generating an altered cellular phenotype which contributes to tissue aging. The two models have many essential similarities.

to the aging process.[5] Over the last thirty years, Hayflick's original observations have been confirmed in literally hundreds of reports involving fibroblasts and have been extended to many renewing or conditionally renewing mammalian cell types. The basic hypothesis that replicative senescence plays a role in organismal aging has also undergone some clarification and refinement over the years. It can now be drawn in a simple flow chart format that captures its essential elements (see Fig. 1). In addition, a number of observations have emerged that are consistent with Hayflick's original contention that cell senescence and aging are associated (see Table 1).

Table 1. Correlative evidence linking replicative senescence and aging

Observations Linking Aging, Age-Related Disease and Cell Senescence	Reference Number
Human cell cultures show a decline in growth potential with advancing donor age.	90-96
The proliferative lifespan of fibroblasts in culture correlates strongly with the lifespan of the species from which they were taken.	97
Caloric restriction extends the lifespan of whole organisms and leads to a reduced number of senescent cells in lens epithelium in vivo.	69
Cultures of vascular endothelial cells derived from atherosclerotic arteries show a greatly reduced lifespan compared to autologous cultures derived from veins.	98
Organ cultured corneae show a strongly age-linked increase in the number of senescent cells in the endothelial layer.	99
The pattern of gene expression in senescent cells in vitro is consistent with the development of age-related degenerative disease in vivo.	100, 101
The number of senescent cells observed within dermal tissue sections increases in an age-dependent manner.	102
The proliferative lifespan of fibroblasts from donors with progeroid syndromes is significantly reduced.	see text

The Kinetics of Replicative Senescence

The original description of in vitro fibroblast growth formulated by Hayflick assumed that the cultures studied were composed of homogeneous populations of cells which were either all growing or all senescent and that the failure to grow resulted from cell death. Both these assumptions, whilst initially sensible, were subsequently shown to be incorrect. Early work on RNA synthesis in growing and senescent fibroblasts showed that tritiated uridine incorporation occurred in all the cells regardless of age. Senescent fibroblasts were thus alive and metabolically active.[7] Extended (72+ hour) pulse labeling experiments performed at every passage with tritiated thymidine and designed to pick up viable cells which never divided showed that unlabelled cells were present in very young cultures and that labeled cells were present in very old cultures.[8] These "cell kinetic" experiments showed that primary cultures in vitro are bimodal mixtures of dividing and senescent cells. The proportions of which alter as the cells are serially passaged. Fraction of labeled mitoses studies, designed to measure the length of cell cycle compartments, also excluded any significant extension of the cell cycle as a cause of the failure of the population to expand.[9] A simplified version of the two kinetic models for cell senescence (and their implications) is shown in Figure 2.

A subsequent series of studies examining the variation in individual clone (or colony) sizes through the lifespans of human fibroblast[10-12] and glial cell[13] cultures provided an explanation for the gradual decline in the labeling index seen in primary human cells. These data demonstrated that primary cultures move from an early state of population expansion in which most of the cells are capable of forming large colonies (and a few are not), through a state in which the population is static or declining as a result of a loss of cell division capacity in most members of the culture although colony forming capacity is retained in a small fraction of cells (see Fig. 3). Studies in human fibroblasts showed that this process does not occur as a result of cultures being mixtures of clones with fixed long and short replicative limits. Rather, daughter cells derived from the same mitosis of a single progenitor cell frequently showed large differ-

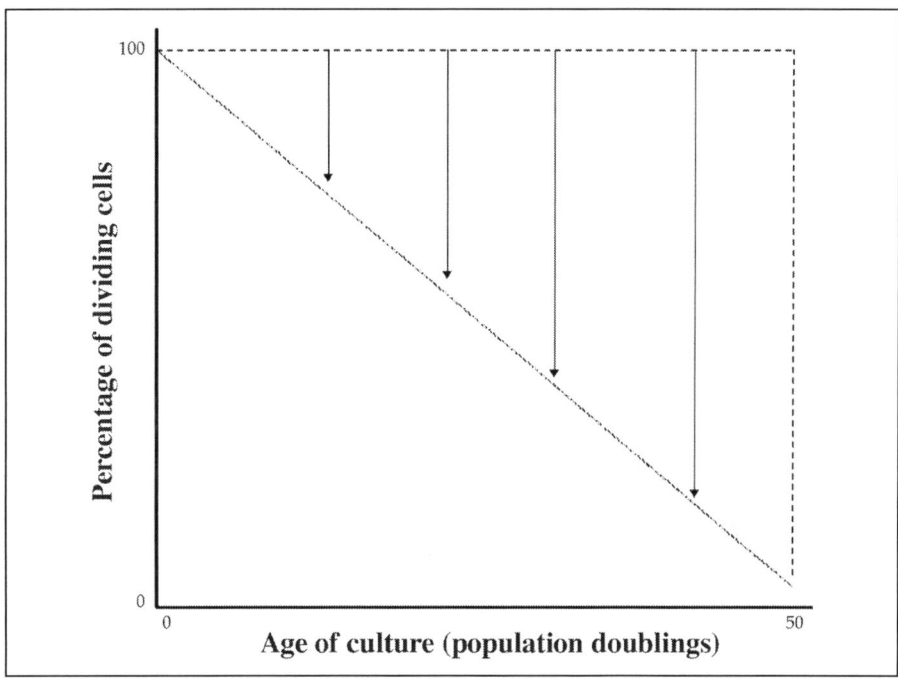

Figure 2. Simple schematic of the difference in divisional fraction between populations displaying unimodal (dashed line) and bimodal kinetics (dotted line). A unimodal population would display a fixed fraction of dividing cells until the very last population doubling when they would enter senescence simultaneously. By contrast, a bimodal population has senescent cells present from the start. Analyses of gene expression within the culture such as Northern blots or microarrays thus give an average for the population and can be misleading unless the fraction of dividing cells is known.

ences in proliferative capacity.[10,14] Taken together, these data are consistent with a counting mechanism that operates at each cell division. This mechanism contains within it a strong stochastic (chance) element. Essentially, at each cell cycle the cell has a chance of committing to senescence or continuing to divide with the relative probability of each event dependent on the number of times the cell has divided before. To date, there is a shortage of detailed kinetic information on cell types other than fibroblasts. This is because such studies tend to be both complex and time consuming. However, the lack of comparative kinetic information (in particular the lack of Pontein plate or Smith-Whitney analysis of colony size changes) is unfortunate because the available evidence shows that both in vitro rates of exit from the cell cycle and baseline apoptotic rates differ between different cell types (even ones that have identical replicative potential).[15,16] These observations have important implications for the rates at which senescent cells are generated.

Cell Kinetics and the Replicative Senescence Hypothesis

A critical observer will note that the central concept of the replicative senescence hypothesis is that of the progressive accumulation of senescent cells as a result of tissue turnover. A reasonable criticism of this postulate is that populations of cells have to undergo large numbers of divisions to generate enough senescent cells to exert an effect. Such divisional limits are unlikely to be routinely encountered in vivo. The observation of reproductive heterogeneity within cell populations rescues the hypothesis from the simple criticism that "you have to do fifty population doublings before any senescent cells turn up" because it implies that senescent cells

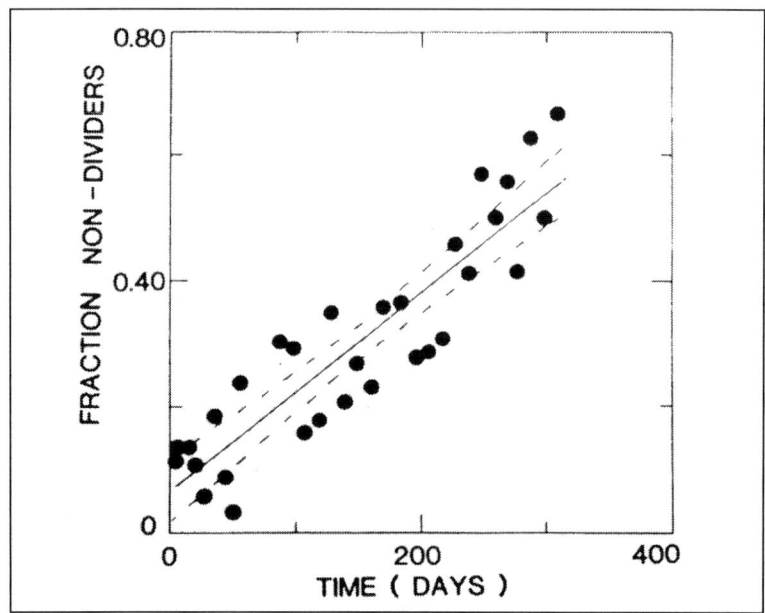

Figure 3. Increase in the senescent fraction of a culture of normal human glial cells (strain787CG) as measured by the palladium-island miniclone technique. In this procedure, the base of a culture dish is covered with agarose (to which cells cannot attach). Then, using a suitable template and an electron microscope shadowing unit, a pattern of palladium is cast onto the agarose. The pattern is that of an outer ring of metal and a clear inner «sea» of agarose broken by a pattern of palladium islands (each island has an area varying from 45,800-18,000 μm^2 depending on the template). Cells are introduced onto the outer ring at normal growth densities and onto the islands such that there is an average of one cell per island. The islands are then monitored for division over a ten day period. This procedure is repeated throughout the replicative lifespan of the culture under analysis until it reaches senescence. The vertical axis is the fraction of islands on which the single founder cell failed to divide in the course of the experiment. The dotted lines show one standard error. Each data point represents n= 1200 individual palladium islands. Adapted from ref. 13.

in vivo will begin to accumulate relatively early. Thus, there is no need for a tissue to "run out of population doublings" before senescent cells can begin to exert their effects.[17] A more thoughtful objection to the whole idea that tissue turnover plays a driving role in aging is to point out that the divisional fraction within normal tissues is frequently vanishingly small; too small in fact for any significant accumulation of senescent cells to occur. Data on the in vivo divisional kinetics of normal tissues (usually obtained by in vivo bromodeoxyuridine (BrdU) labeling studies) are initially supportive of this idea (see Table 2). However, it is often forgotten that these small amounts of daily cell division are taking place over very long periods of time. If likely organismal lifespans (rather than likely experimental duration) are taken as the frame of reference, the amounts of cell turnover that tissues probably undergo are in fact quite substantial (see Table 2). Use of the word "substantial" highlights an unsolved problem with the replicative senescence hypothesis of aging (and, to my mind, all current mechanistic hypotheses of aging). At the time of writing, we have no clear idea of the precise number of senescent cells (or aberrant mitochondria or oxidized protein) required to induce a physiological deficit in any tissue (although it is likely to be not so much a function of absolute number as it is one of relative frequency and distribution). Within the replicative senescence field this is due to a simple lack of good, unambiguous markers (antisera for choice) that can be applied to the systematic morphometry of normal and diseased tissue from an age range of donors.

Table 2. Approximate estimates of cell turnover in vivo during the course of the lifetime of several species based on single time point labelled DNA precursor studies

Cell Type	Labelling Index(%)[1]	Estimated Population Doublings in Lifespan[2]	Reference
Mouse Forebrain Glia	0.2	6.4	103
Mouse Liver	0.4	13	103
Mouse vascular endothelium	0.4	13	104
Human corneal endothelium	0.001		105

[1] Based on in vivo (or immediate ex vivo) labelling studies with ^{3}H-thymidine or bromodeoxyuridine.
[2] Based on estimated mouse lifespan of two years and human lifespan of 80 years calculated using the formula. $(T)=\lambda(T_s)/ L.I^{104}$ where:
T is the turnover time (population doubling time) in hours
T_s is the duration of S phase (set at an arbitrary 8 hours)
γ is a factor for slowly proliferating cells with a value close to $\log_e 2$
L.I. is the absolute (not percentile) labelling index.
N.B. these are not absolute figures, they are essentially "back of the envelope" calculations.

As an aside, it is worthwhile pointing out that a significant body of work exists on a phenomenon that I have previously termed "reactive" cell senescence.[18] This is the immediate entry by members of a population of cells into senescence as a result of an external stimulus (such as treatment with γ-radiation or infection with vectors carrying activated viral oncogenes such as RAS-V12).[19-20] Conceptually, the contribution that reactive senescence may make to aging is very similar to that advanced by Cutler's dysdifferentiation hypotheses (which proposes, in essence, that an extrinsic stress triggers a dramatic shift in gene expression affecting both the cell and its surrounding microenvironment).[21] The replicative senescence hypothesis postulates that the senescent cell itself (rather than any particular "replicometer" or mode of production) is a primary effector mechanism in mitotic tissue aging. Thus, the observation of reactive senescence gives another plausible route by which senescent cells could be generated in vivo and adds to the likely relevance of cell senescence to organismal aging. If drink is the primary effector mechanism in drunkenness, the people who bring it to the party are less important than the fact it's on the table.

The Potential Impact of Cellular Senescence on Human Tissues

Once senescent, human cells display a variety of characteristics. Firstly, a cessation of DNA replication under normal conditions. The molecules directly responsible for this process were gradually characterized (largely via cell fusion experiments) as the cyclin dependent kinase inhibitors. The principle molecule involved in bringing about senescence in fibroblasts appears to be p21^{waf} a broad-spectrum inhibitor of the cyclin-CDK complexes (CDK: cyclin dependent kinase) with a secondary role being played by p16^{INk4a} inhibition of CDK4-cyclin D kinase activity. Secondly, and most importantly in my view, senescent cells show an altered spectrum of gene expression. It should be noted that senescence is a highly active but selective process. Some genes are repressed and some genes are upregulated by a variety of mechanisms including both increased transcription and increased mRNA stabilization. This is perhaps best illustrated with respect to the *c-fos* proto-oncogene. Repression of *c-fos* is a characteristic event in the replicative failure of T cells[22] and fibroblasts[23] (although it does not occur in senescent melanocytes). However, a number of other cell cycle associated genes (including *c-myc* and *H-ras*) are not repressed.[24] Table 1 lists a number of the principle changes identified in senescent cells. Many more have been reported as a result of initial microarray studies.[25]

One route by which such cells could cause a reduction in the functional efficiency of a mitotic tissue is simply via impairment of its regenerative potential. This could result from the presence of significant numbers of fully senescent (and thus division incompetent) cells, an overall approach of cells towards (but not at) the end of their lifespan, or both. The prior cell division necessary to reach this state could reflect the tissue having relatively high routine cell turnover, or might result from localized bursts of cell division in response to damage or infection.

The immune system illustrates the potential problems caused by senescent cells particularly well. The demonstration of the senescence of T lymphocyte cultures overturned a prevailing preconception among immunologists that normal T lymphocytes were immortal. A decline in T cell proliferative response in vivo has been reported with aging and correlates with an increasing fraction of non-mitogen responsive cells and a fraction of G2-arrested cells with similar characteristics to those of restimulated cultures of senescent T cells.[26-28] There is also evidence for a population of "presenescent" T cells with reduced division potential in aged individuals.[28] If the proliferative potential of T cells is the same within the body as it is ex vivo, then the number of expansions which a particular cell clone can undergo in response to antigen stimulation is not indefinite and is probably less than ten.[29] In addition to having a limited capacity to respond, themselves, senescent cells "take up space" within the peripheral T cell pool (which maintains an effectively constant number of T cells over time).[30,31] Thus, senescent T cells have the potential to reduce the rate of naive T cells entering the peripheral pool from the thymus. Given the emerging data suggesting that many tissues in the body appear to have a secondary regeneration capacity based (at least in part) on the capture of circulating progenitor cell populations,[32] it will be of considerable interest to determine if senescent cells impede tissue regeneration by failing to "make space" for progenitor cell uptake.

The potential effect of senescent cells is not limited to a reduction in division potential. The presence of senescent cells could also have an effect, as often they over-express proteins that act at a distance, such as the classic example of collagenase over-expression by senescent dermal fibroblasts.[33,34] There is also some evidence that the presence of senescent cells alters the behavior of cells surrounding them with deleterious consequences. Evidence of this is provided by the observation that under normal conditions, plasminogen activator inhibitor-1 (PAI-1) activity in endothelial cells is down regulated by factors produced by attendant smooth muscle cells or fibroblasts.[35] However, if senescent cells of either type are substituted for their young counterparts, the down-regulation of PAI-1 activity either disappears or is replaced by a significant up-regulation. This is a potentially serious phenotypic shift since elevated PAI levels in vivo are a major risk factor for myocardial infarction and deep vein thrombosis. Accordingly, transgenic studies have shown that PAI-1 over-expression leads to thrombotic disease.[36,37]

Do Cells Count Divisions and if so How?

The kinetic analysis of human fibroblasts long suggested that if such cells in fact possessed a programmed mechanism for counting cell divisions the "replicometer" was not a simple timing mechanism or "tally stick". It had to contain a substantial stochastic component. A number of mechanisms (including progressive loss of methylation)[38] were proposed as potential stochastic replicometers in the 1970s and 1980s but this chapter will focus on the merits of only two. These are (i) the hypothesis that divisional history could be monitored, and replicative senescence triggered, by the progressive loss of sequences from the ends of chromosomes (telomeres)[39,40] and (ii) the proposal that increasing concentrations of a long lived inhibitory protein could act as an effective divisional timer limiting replicative lifespan.[41]

Of these mechanisms, telomeric attrition is by far the best established and most convincing. Human chromosomes are linear pieces of DNA and telomeres distinguish such natural chromosome ends from simple double-strand breaks (which can lead to recombination between chromosomes and the activation of genome damage-monitoring systems). Telomeres enable the ends of human chromosomes to behave so differently from simple breaks because they are arrays of a specific terminal DNA sequence (the hexamer TTAGGG). Telomeric sequence acts

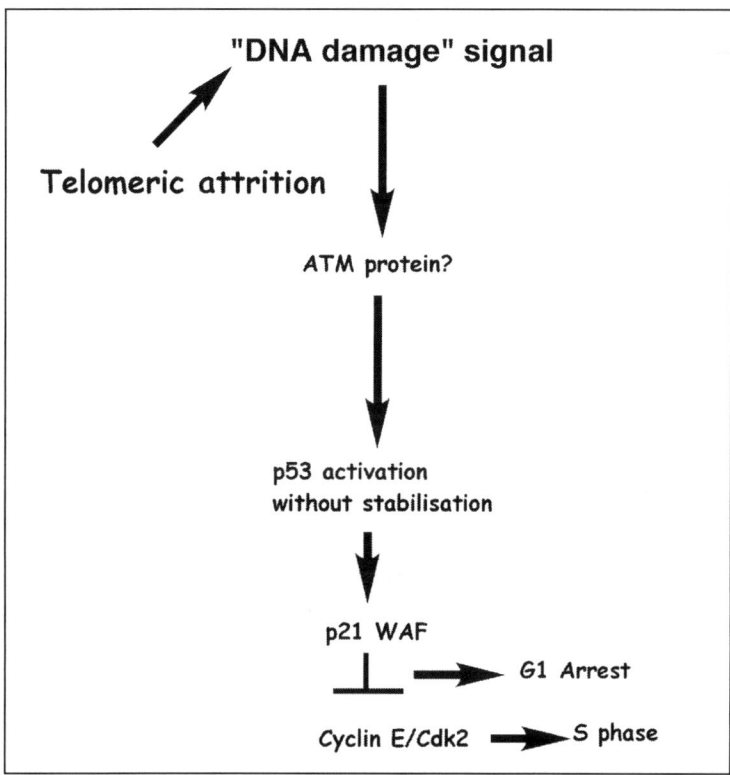

Figure 4. Telomere-driven senescence. Progressive telomeric attrition leads to the generation of one or more «critically short» telomeres that signal cell cycle arrest via the activation of the p53 tumor supressor gene product. p53 is a transcription factor for a number of genes, among them, the cyclin dependent kinase inhibitor p21^{waf}. Expression of p21 leads to cell cycle arrest at the G1-S phase boundary. ATM is the Ataxia telangiectasia mutated gene product. Adapted from ref. 61.

as a binding site for a wide range of proteins including hTRF1 and hTRF2 which act together to form a complex structure known as the T loop.[42,43]

In the absence of any mechanism to produce compensatory de novo synthesis of telomeric DNA, the chromosomes face what has been termed the "end replication problem". This results from the inability of all known DNA-dependent DNA polymerases to commence synthesis de novo. At the very termini, a region at least as large as the priming RNA primer for lagging strand DNA synthesis will be systematically deleted every time the cell undergoes division. In contrast, in most immortal human cell lines there appears to be compensatory de novo synthesis of telomeric DNA by the enzyme telomere reverse transcriptase (telomerase) or at least mechanisms that produce a relatively stable maintenance of telomere length (the alternative lengthening of telomeres or ALT pathway).[44,45] Progressive loss of telomeres, with the exact rate being a function of the cell type under study and the degree of oxidative damage present at the telomere, leads to the inability of a telomere to maintain T loop integrity.[46] This triggers the activation of the p53 protein,[47,48] transcription of p21^{waf} and cell cycle arrest at the G1/S phase transition of the cell cycle (see Fig. 4). A particularly attractive feature of telomere shortening as a stochastic counting is the fact that end replication loss is inherently asymmetric between both strands of the double helix (leading to relatively "long" and "short" strands). Mathematical modeling studies essentially based simply on the random inheritance of the long and short strands of the various chromosomes between daughter cells predict a very similar colony size distribution to that actually observed in human fibroblasts.[49-51]

The cloning of human telomerase (hTERT) allowed the hypothesis that replicative senescence could be "telomere-driven" to be tested by direct intervention means.[52] Ectopic expression of telomerase is now known reproducibly to immortalize a wide range of normal human cell types.[52,53] Immortalization using hTERT differs significantly from that observed with SV40 large T antigen (the previous oncogene of choice for the construction of cell lines). SV40 immortalization requires secondary events at mutational frequency to bypass "crisis".[54] Telomerase immortalization does not and the frequency of immortalization is thus approximately six orders of magnitude higher, suggesting that a primary lifespan control system is being interdicted.[55] This, together with a large body of data showing the relatively normal phenotypic performance of hTERT-immortalized cells in a range of assays, strongly suggests that many human cell types do use some function of telomere length to monitor divisions and trigger senescence.[46,56]

However, a number of reports now show that not all human cell types will immortalize in vitro with telomerase alone.[57] Human keratinocytes require telomerase and either the E7 gene product or tissue culture conditions that down regulate the expression of $p16^{INK4a}$ to become immortal.[58,59] Similar observations have been reported for pancreatic β-cells[60] and my own laboratory has recently found that human corneal endothelium will not immortalize in the presence of ectopic telomerase alone but will do so if the hTERT gene is co-transfected with the human CDK4 gene (unpublished observations). Taken together, these observations suggest the existence of a telomere-independent growth arrest pathway centered on the retinoblastoma-$p16^{INK4a}$ axis[61] (Fig. 5). There is limited theoretical support for this pathway as a cell division counting system. P16 is both a cell cycle inhibitory protein and one with an extremely long half-life. These characteristics were predicted many years ago as requirements for a simple "protein driven" molecular model which would give rise to the kinetic behavior seen in fibroblast cultures. This model predicts the gradual accumulation of a Mortallization (or M) protein, which competes with a Division (or D) protein, produced at a constant concentration, for binding sites upstream of a gene coding for an initiator of DNA synthesis. Binding of the D protein permits initiator synthesis and cell cycle traverse. However, if the M protein binds, it prevents binding of the D protein, irreversibly blocks production of the initiator and inhibits cell division.[62] A stochastic element is built into this model by the relative concentrations of M and D and their competition for binding. The competition principles underlying this model hold for the binding of p16 to its target cyclin-CDK complexes and their interaction with their targets.

The existence of telomere-independent senescence (or TIS) is somewhat controversial. It has been suggested that it simply reflects inadequate tissue culture conditions, a view which is lent some support by the observation that p16 accumulation can be modulated by changes to tissue culture practice (such as growth on feeder layers[58] or alterations to the partial pressure of oxygen in the incubator).[63,64] Personally, I consider this to be possible, but unsupported by experimental evidence at the current time. In order for human telomere independent senescence to be an in vitro artifact, it would have to be absent in all tissues in vivo at all times. At least it would have be nominally absent in vivo from the cell types in which it has been reported in vitro. To demonstrate this would require a comparative morphometric analysis using markers that distinguish between cells that have undergone telomere-driven senescence and TIS. The same type of analysis would be necessary just to demonstrate that the fraction of cells undergoing TIS is so small compared to the fraction of cells undergoing telomere-dependent senescence that the proportion of cells undergoing TIS in vitro is artifactually large. One might even wish to distinguish between cells that have entered TIS as a result of divisional arrest via the p16-Rb pathway and those that have arrested by "reactive" senescence mechanisms (such as ceramide-induced arrest).[65] Given that there is a current lack of good markers that can be used in morphometry to detect any senescent cells in tissue, it seems speculative at best (and narrow

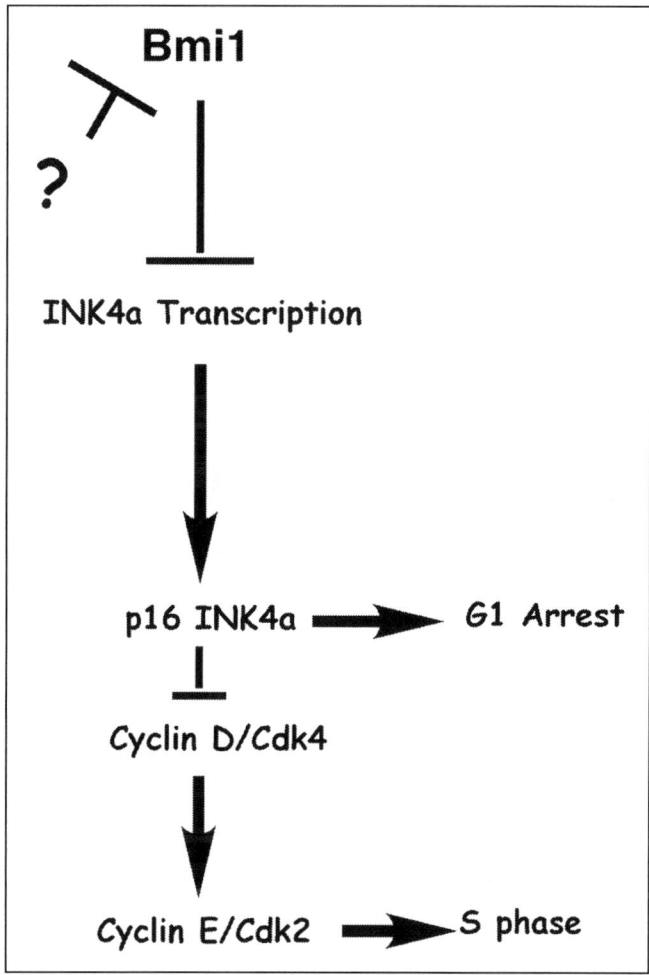

Figure 5. Simple schematic telomere-indendent induction of the senescent state. The product of the polycomb gene *Bmi-1* repressed the transcription of the cyclin dependent kinase inhibitor p16^{INK4a}. Derepression at this locus leads to a build up of p16 protein and inhibition of the activity of cyclin D-Cdk4 and cyclin E-Cdk2 kinase pairs. This leads to a failure to phosphorylate the retinoblastoma gene product and a failure to pass the G1-S phase boundary. Adapted from ref. 61.

minded at worst) to dismiss in vivo human TIS, at least for the time being. In vitro experiments designed to detect an actual counting mechanism for TIS rather than simple random events forcing cell cycle exit in a fraction of the population are fraught with difficulties of interpretation.

It seems clear that at least one class of mammal lacks telomere-driven senescence altogether. Rodents have very long telomeres relative to humans. Primary and cultures of rodent fibroblasts frequently become senescent whilst still telomerase positive.[66,67] The whole process of rodent fibroblast senescence is also far less efficient in mice than in men (spontaneous escape frequencies from senescence of ~1 in 10^{-6} per cell per generation are conventionally quoted for rodent fibroblasts compared to ~1 in 10^{-12} for human fibroblasts). In addition, unlike human fibroblasts, those from rodents arrest with p53 both active and stable. Taken together, these data suggest that rodent fibroblast senescence is under different genetic controls to its human

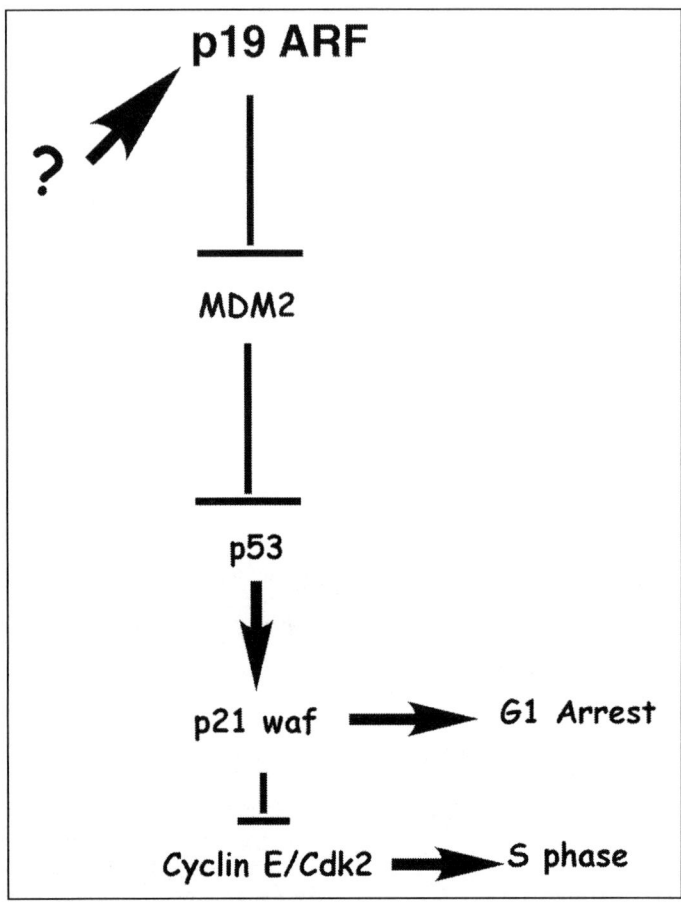

Figure 6. Simplified pathway by which rodent fibroblasts may enter replicative senescence. Upregulation occurs at the ARF (alternate reading frame) locus in response to a variety of upstream signals including v-Ras and v-Abl. ARF neutralizes the ability of Mdm2 to promote p53 degradation, leading to the stabilisation and accumulation of p53. Ectopic expression of ARF causes a senescence-like growth arrest. This pathway is p53 dependent but telomere independent. Adapted from refs. 61 and 89.

counterpart (Fig. 6). The apparent lack of telomere-driven senescence in rodents has again produced an element of controversy. It has been suggested that senescence in rodent fibroblasts in vitro is essentially artifactual and results from "culture shock" as a result of poor tissue culture conditions including, but not restricted to, hyperoxia.[68] Whilst there is probably a significant element of truth in this, it is a very easy position to overstate (particularly because the cliché is so catchy). The real points at issue may be summarized as (i) do rodent fibroblasts show cell senescence in vivo at all? And (ii) if so do they have a counting mechanism for cell divisions in vivo or is all rodent senescence essentially "reactive"? These questions may have fundamental implications for the phenotype of the Wrn knockout mouse.

My personal view is that rodent fibroblasts certainly have senescent cells in vivo and may have counting mechanisms that can operate both in vitro and in vivo. I base this statement on two pieces of data. Firstly, combinations of long BrdU labels and colony forming studies conducted using the lens epithelium of young old and calorie restricted mice show a gradual decline in both the number of lens cells that can incorporate label in vivo and the in vitro colony

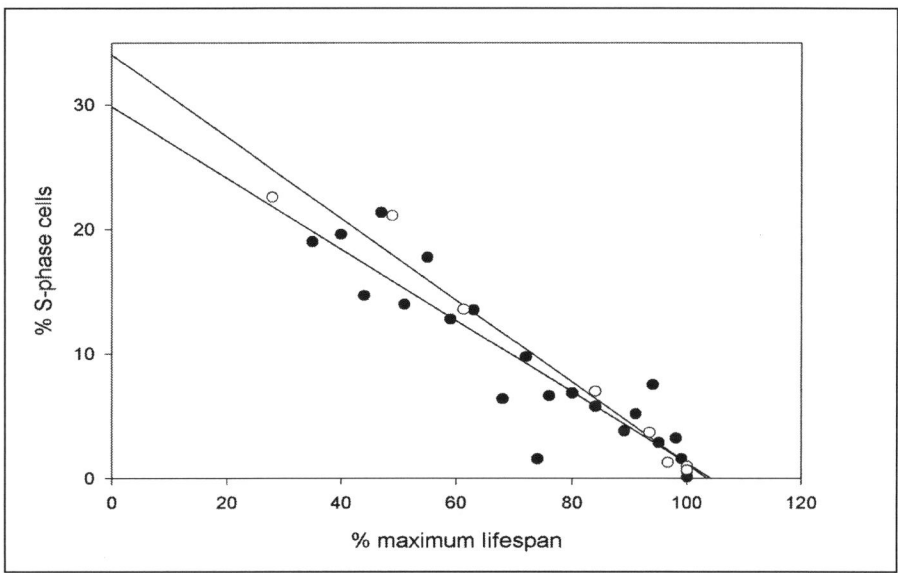

Figure 7. Rate of loss of the S phase fraction in cultures of primary rat fibroblasts (open dots) and primary human fibroblasts (closed dots). Note the similarity of the decline rates as a function of maximum replicative lifespan. Data reproduced from ref. 90.

forming ability of such cells.[69] Many lens epithelial cells from the aged animals fail to form colonies in a Smith-Whitney style cloning assay and thus meet the in vitro definition of senescent. These data are consistent with the presence of senescent cells in rodents in vivo. Secondly, it is known that the kinetics of exit of rodent fibroblasts from the cell cycle are very similar to those observed for normal human fibroblasts (Fig. 7). If studied by mini-cloning, cultures of rodent fibroblasts start off with a large number of clones with good reproductive capacity and gradually shift to a state in which most of the cells are non-dividing but a few still form colonies[70,71] (the high spontaneous immortalization rate in rodents renders these studies difficult, but not impossible to perform). This is an observation that I personally find easier to reconcile with the presence of a divisional counting mechanism than random arrest due to oxidative stress because it would appear to require large variations in the antioxidant capacity of clonal cells of the same type from the same donor. This being said, there are a significant number of practical experimental factors that could give such a range of colony size formation as a result of "culture shock": (such as prior cell history in the body, different microniches within the tissue culture dish etc). Unfortunately, experiments that can unambiguously distinguish between programmed exit from the cell cycle and arrest as a result of random damage are very difficult to design. All that can be said at the moment, in my view, is that no experiment has disproved the existence of programmed cell division counting in rodents. It might thus be premature to dismiss its existence out of hand.

Werner's Syndrome: Putting It All Together

Two striking cellular phenotypes arise as a result of loss of function mutations in the *WRN* gene. The first of these (and one of the clearest features of Werner's syndrome) is the extremely poor replicative potential of cultured fibroblasts. Literature comparisons of the lifespans of all Werner's syndrome cell strains published with those of published normal controls, show that 90% of Werner's syndrome cultures have an in vitro lifespan of less than 20 population doublings.[72] The cause of this limited replicative lifespan is a greatly increased rate of decline in the mitotic fraction of Werner's syndrome fibroblasts (as measured using either short BrdUrd pulse labels or the expression of endogenous markers of proliferation such as PCNA or pKi67).[73,74]

The second distinct property of WS cells is a mutator phenotype that is most readily demonstrated by selection experiments for *HPRT* mutants using 6-thioguanine (6-TG).[75,76] WS cultures produce a significantly higher fraction of 6-TG resistant colonies than wild type controls most of which are caused by large deletions. A hyper recombinogenic phenotype has also been demonstrated in fibroblasts using plasmids containing overlapping fragments of the neomycin gene. Related studies in which linearized plasmids were introduced into WS lymphoblastoid cell lines have also shown that a significant number of rejoining events result in deletion-mediated mutations.[77,78] How do these cellular manifestations of the disease relate to one another, to the clinical presentation of the patient and to the aging process in general? Three possibilities suggest themselves as answers to this question. (i) That Werner's syndrome is of no real value in understanding normal aging and that studies of the pathology of Werner's syndrome sheds light only on the pathology of Werner's syndrome. (ii) That the key feature of Werner's syndrome is the genomic instability caused by loss of WRN and Werner's syndrome may thus inform significantly on the relationship between normal aging and "DNA damage". (iii) That the premature replicative senescence of Werner's syndrome is the key feature of the disease and the resultant pathology can provide valuable data on the role played by senescent cells (or reduced proliferative capacity) in normal aging.

In accordance with this logic, ectopic expression of telomerase (hTERT) was forced in WS fibroblasts in order to determine if their premature senescence resulted from global genomic instability or was due to deletions occurring at or near the telomere as a result of the absence of WRN.[79] Based on the analysis of four strains, it was known, at the time these studies were initiated, that the rate of telomere shortening appeared to be greater in WS fibroblasts. However, the WS strains entered senescence at a telomere lengths ~0.7 kb longer than those seen in normal controls.[80] These observations led to the suggestion that accelerated telomere shortening was unlikely to be the cause of the higher rate of senescence of WS fibroblasts. However, the study in question measured mean terminal restriction fragment length, not true telomere length and this is known to mask population heterogeneity in telomere length at senescence (e.g., an altered distribution of telomere lengths rather than simply a higher mean). In fact WS fibroblasts immortalize normally following the reintroduction of telomerase.[79] A finding which indicates that the primary cause of premature replicative senescence in the disease results from altered telomere loss dynamics arising from the absence of WRN. Loss of WRN protein thus causes an acceleration of normal telomere-driven senescence. Telomere-driven senescence is indeed accelerated in Werner's syndrome. Hence, tissues containing telomere-driven cell types would be expected to show degenerative changes if replicative senescence played a causal role in aging.[81] T cells are known to display such telomere-dependent senescence but Werner's syndrome patients display essentially normal immune function.[82] However, it is known that mass T cell cultures derived from the patients display no lifespan deficit compared to those taken from normal controls.[83] Since Werner's syndrome T and B cells display elevated mutation rates at the *HPRT* locus following selection with 6-TG this again appears to argue against any effect on replicative lifespan as a result of genomic instability in any significant fraction of the T cell clones.[84] The observation of significant markers of global genomic damage in tissues which are both clinically affected and clinically unaffected thus appears inconsistent with the hypothesis that genomic instability is a primary driver in mitotic tissue aging in Werner's syndrome. In contrast, the appearance of premature replicative senescence in an affected tissue (the dermis) but not in an unaffected one (the immune system) is consistent with a causal role for senescent cells in the development of the progeroid pathology seen in the disease.

Elizabeth Ostler and I recently formulated and published a model[85,86] in which we attempted to convey the general principles underlying these studies and to suggest further work. A line of reasoning with many essential similarities to our model was also recently put forward

independently by Johnson and colleagues.[87] Essentially the model is based on the following postulates.
  i. The available data on Werner's syndrome are consistent with a true "acceleration" of replicative cell aging in a selected series of tissue lineages.
  ii. The observation that some tissues are severely affected in Werner's syndrome but some are clinically unaffected provides us with a valuable way to understand the impact that cell senescence has on tissues.
  iii. Observation of an abbreviated replicative lifespan in the absence of any accelerated tissue aging will provide potent evidence against the hypothesis that cell senescence plays a major role in the aging of that tissue. Conversely, the demonstration that normal rates of tissue aging in Werner's syndrome correlate with the absence of premature cell senescence will greatly strengthen the hypothesis that cell senescence plays a causal role in the aging of mitotic tissue. This renders the model subject to falsification and thus (hopefully) makes it useful.
  iv. Different cell types may well use different counting mechanisms to monitor divisional history (this holds for both different tissues compared in the same human and the same tissue compared between different species).

Although a full analysis of many of the most interesting cell types in Werner's syndrome has not yet been undertaken, it is possible to propose a general model of what is likely to be found. On the basis of the data currently available we predict that:
  i. Truly post-mitotic cells will be unaffected by *WRN* mutations, since the phenotypic impact of the disease is based upon the generation of senescent cells. This emphasizes the limits of the replicative senescence hypothesis which is formulated to explain the aging of mitotic tissue.
  ii. Tissues that normally lack expression of the *WRN* gene should be unaffected in a patient with the disease (or at least be affected only as a result of "cross-talk" with an affected tissue).
  iii. Tissues that normally show telomere-independent senescence should be unaffected. The corollary of this prediction is that tissue types unaffected in Werner's syndrome may be prime candidates for telomere-independent normal senescence mechanisms. It is of course possible that the accelerated telomeric loss caused by the inactive WRN may "flip" a cell type from a telomere-independent to a telomere-dependent state, with subtle phenotypic consequences.
  iv. Tissues that show small end-replication losses (such as fibroblasts) should be severely affected in Werner's syndrome. However, the amount of DNA lost as a result of the end-replication problem varies between cell types (as a result of variations in the length of the 3' overhang). A priori, loss of *WRN* should impose a fairly fixed additional rate of telomeric loss (based on a loosely fixed frequency of replication fork stalls in response to adducts). Thus, as normal telomeric loss rate increases, the additional loss caused by a mutation in *WRN* probably becomes progressively less significant. The replicative capacity of tissues with a high endogenous telomeric loss rate is, therefore, likely to be only marginally decreased by loss of the Werner's syndrome helicase.

Conclusion

Although significant correlative experimental data is available for the hypothesis that replicative senescence contributes to human aging, direct interventional falsification of it (e.g., by the deliberate acceleration of replicative senescence in an organism in the expectation that this will lead to the acceleration of mitotic tissue aging) remains just outside the practical capacity of the gerontological community at the current time. Given this situation, a great advantage of Werner's syndrome is that its clinical features, in particular the fact that some tissues are severely affected whilst others are mildly affected (if at all) presents a clear opportunity to study the relationships between senescence, aging and disease in the whole organism. The existence of mouse knockouts for the Werner's syndrome homologue now provide a valuable additional resource. However, the data obtained in mouse studies must be evaluated in the context of the differences in the biology of replicative senescence between humans and rodents.

References

1. Martin GM. Genetics and ageing: The Werner syndrome as a segmental progeroid syndrome. Ad Exp Biol Med 1985; 190:161-171.
2. Popper K. Conjectures and refutations: The growth of scientific knowledge. Routledge Classics Series. London & New York.
3. Carrel A. Artificial activation of the growth in vitro of connective tissue. J Exp Med 1913; 16:14-19.
4. Hayflick L, Moorhead PS. The serial cultivation of human diploid fibroblast cell strains. Exp Cell Res 1961; 25:585-621.
5. Hayflick L. The limited in vitro lifespan of human diploid cell strains. Exp Cell Res 1965; 37:614-636.
6. Hayflick L. The cell biology of aging. J Invest Dermatol 1979; 73:8-14.
7. Maciera-Coelho A, Ponten J, Philipson L. The division cycle and RNA synthesis in diploid human cells at different passage levels in vitro. Exp Cell Res 1966; 42:673-684.
8. Cristofalo VJ, Scharf BB. Cellular senescence and DNA synthesis. Exp Cell Res 1973; 76: 419-427.
9. Grove GL, Cristofalo VJ. Characterization of the cell cycle of cultured human diploid cells: effects of aging and hydrocortisone. J Cell Physiol 1977; 90:415-422.
10. Smith JR, Whitney RG. Intraclonal variation in proliferative potential of human diploid fibroblasts: stochastic mechanism for cellular ageing. Science 1980; 207:82-84.
11. Smith JR, Hayflick L. Variation in the life-span of clones derived from human diploid cell strains. J Cell Biol 1974; 62:48-53.
12. Smith JR, Pereira-Smith OM, Schneider EL. Colony size distributions as a measure of in vivo and in vitro aging. Proc Natl Acad Sci USA 1978; 75:1353-1356.
13. Ponten J, Stein WD, Shall S. A quantitative analysis of the ageing of human glial cells in culture. J Cell Physiol 1983; 117:342-352.
14. Jones RB, Whitney RG, Smith JR. Intramitotic variation in proliferative potential: stochastic events in cellular aging. Mech Ageing Dev 1985; 29:143-149.
15. Kalashnik L, Bridgeman CJ, King AR et al. A cell kinetic analysis of human umbilical vein endothelial cells. Mech Ageing Dev 2000; 120:23-32.
16. Thomas E, al-Baker E, Dropcova S et al. Different kinetics of senescence in human fibroblasts and peritoneal mesothelial cells. Exp Cell Res 1997; 236:35535-8.
17. Faragher RGA. Cell senescence and human aging: Where's the link? Biochem Soc Trans 2000; 28:221-226.
18. Faragher RG, Mulholland B, Tuft SJ et al. Aging and the cornea. Br J Ophthalmol 1997; 81:814-817.
19. Serrano M, Lin AW, McCurrach ME et al. Oncogenic ras provokes premature cell senescence associated with accumulation of p53 and p16INK4a. Cell 1997; 88:593-602.
20. Sewing A, Wiseman B, Lloyd AC et al. High-intensity Raf signal causes cell cycle arrest mediated by p21Cip1. Mol Cell Biol 1997; 17:5588-5597.
21. Zs-Nagy I, Cutler RG, Semsei I. Dysdifferentiation hypothesis of aging and cancer: a comparison with the membrane hypothesis of aging. Ann NY Acad Sci 1988; 521:215-225.
22. Sikora E, Kaminska B, Radziszewska E et al. Loss of transcription factor AP1 DNA binding activity during lymphocyte aging in vivo. FEBS lett 1992; 312:179-182.
23. Sheshadri T, Campisi J. Repression of c-fos and an altered genetic programme in senescent human fibroblasts. Science 1990; 247:205-209.
24. Rittling SR, Brooks KM, Cristofalo VJ et al. Expression of cell cycle dependent genes in young and senescent WI38 fibroblasts. Proc Natl Acad Sci USA 1986; 83:3316-3320.
25. Shelton DN, Chang E, Whittier PS et al. Microarray analysis of replicative senescence. Curr Biol 1999; 9:939-945.
26. Perillo NL, Naeim F, Walford RL et al. In vitro cellular aging in T lymphocyte cultures: analysis of DNA content and cell size. Exp. Cell Res 1993; 207:131-135.
27. Kubbies M, Schindler D, Hoehn H et al. BrdU-Hoechst flow cytometry reveals regulation of human lymphocyte growth by donor-age-related growth fraction and transition rate. J Cell Physiol 1985; 125:229-234.
28. Inkeles B, Innes JB, Kuntz MM et al. Immunological studies of Aging. III. Cytokinetic basis for the impaired response of lymphocytes from aged humans to plant lectins. J Exp Med 1977; 145:1176-1187.

29. Effros RB, Pawelec G. Replicative senescence of T cells: Does the Hayflick Limit lead to immune exhaustion? Immunol Today 1997; 18:450-454.
30. Aspinall R. Longevity and the immune response. Biogerontology 2000; 1:273-278.
31. Andrew D, Aspinall RJ. Il-7 and not stem cell factor reverses both the increase in apoptosis and the decline in thymopoiesis seen in aged mice. Immunol 2001; 166:1524-1530.
32. LaBarge MA, Blau HM. Biological progression from adult bone marrow to mononucleate muscle stem cell to multinucleate muscle fiber in response to injury. Cell 2002; 111:589-601.
33. Funk WD, Wang CK, Shelton DN et al. Telomerase expression restores dermal integrity to in vitro-aged fibroblasts in a reconstituted skin model. Exp Cell Res 2000; 258:270-278.
34. West MD, Pereira-Smith OM, Smith JR. Replicative senescence of human skin fibroblasts correlates with a loss of regulation and overexpression of collagenase activity. Exp Cell Res 1989; 184:138-147.
35. Zhang JC, Fabry A, Paucz L et al. Human fibroblasts downregulate plasminogen activator inhibitor type-1 in cultured human macrovascular and microvascular endothelial cells. Blood 1996; 88:3880-3886.
36. Auwerx J, Bouillon R, Collen D et al. Tissue-type plasminogen activator antigen and plasminogen activator inhibitor in diabetes mellitus. Arteriosclerosis 1988; 8:68-72.
37. Erickson LA, Fici GJ, Lund JE et al. Development of venous occlusions in mice transgenic for the plasminogen activator inhibitor-1 gene. Nature 1990; 346:74-76.
38. Holliday R. Strong effects of 5-azacytidine on the in vitro lifespan of human diploid fibroblasts. Exp Cell Res 1986; 166:543-552.
39. Wright WE, Shay JW. Cellular senescence as a tumor-protection mechanism: The essential role of counting. Curr Opin Genet Dev 2001; 11:98-103.
40. Olovnikov AM. Telomeres, telomerase, and aging: Origin of the theory. Exp Gerontol 1996; 31:443-448.
41. Shall S, Stein WD. A mortalization theory for the control of the cell proliferation and for the origin of immortal cell lines. J Theor Biol 1979; 76:219-231.
42. de Lange T. Protection of mammalian telomeres. Oncogene 2002; 21:532-540.
43. Smogorzewska A, van Steensel B, Bianchi A et al. Control of human telomere length by TRF1 and TRF2. Mol Cell Biol 2000; 20:1659-1668.
44. Newbold RF. The significance of telomerase activation and cellular immortalization in human cancer. Mutagenesis 2002; 17:539-550.
45. Grobelny JV, Kulp-McEliece M, Broccoli D. Effects of reconstitution of telomerase activity on telomere maintenance by the alternative lengthening of telomeres (ALT) pathway. Hum Mol Genet 2001; 10:1953-1961.
46. von Zglinicki T. Telomeres and replicative senescence: Is it only length that counts? Cancer Lett 2001; 168:111-116.
47. Gire V, Wynford-Thomas D. Reinitiation of DNA synthesis and cell division in senescent human fibroblasts by microinjection of anti-p53 antibodies. Mol Cell Biol 1998; 18:1611-1621.
48. Ma Y, Prigent SA, Born TL et al. Microinjection of anti-p21 antibodies induces senescent Hs68 human fibroblasts to synthesize DNA but not to divide. Cancer Res 1999; 59:5341-5348.
49. Tan Z. Intramitotic and intraclonal variation in proliferative potential of human diploid cells: explained by telomere shortening. J Theor Biol 1999; 198:259-268.
50. Rubelj I, Vondracek Z. Stochastic mechanism of cellular aging—abrupt telomere shortening as a model for stochastic nature of cellular aging. J Theor Biol 1999; 197:425-438.
51. Levy MZ, Allsopp RC, Futcher AB et al. Telomere endreplication problem and cell aging. J Mol Biol 1992; 225:951-960.
52. Bodnar AG, Ouellette M, Frolkis M, et al. Extension of life-span by introduction of telomerase into normal human cells. Science 1998; 279:349-352.
53. Condon J, Yin S, Mayhew B et al. Telomerase immortalization of human myometrial cells. Biol Reprod 2002; 67:506-514.
54. Wright WE, Shay JW. The two-stage mechanism controlling cellular senescence and immortalization. Exp Gerontol 1992; 27:383-389.
55. Shay JW, Wright WE. Quantitation of the frequency of immortalization of normal human diploid fibroblasts by SV40 large T-antigen. Exp Cell Res 1989; 184:109-118.
56. Martens UM, Chavez EA, Poon SS et al. Accumulation of short telomeres in human fibroblasts prior to replicative senescence. Exp Cell Res 2000; 256:291-299.
57. Rheinwald JG, Hahn WC, Ramsey MR et al. A two-stage, p16INK4A- and p53-dependent keratinocyte senescence mechanism that limits replicative potential independent of telomere status. Mol Cell Biol. 2002; 22:5157-5172.

58. Ramirez RD, Morales CP, Herbert BS et al. Putative telomere-independent mechanisms of replicative aging reflect inadequate growth conditions. Genes Dev 2001; 15:398-403.
59. Kiyono T, Foster SA, Koop JI et al. Both Rb/p16INK4a inactivation and telomerase activity are required to immortalize human epithelial cells. Nature 1998; 396:84-88.
60. Halvorsen TL, Beattie GM, Lopez AD et al. Accelerated telomere shortening and senescence in human pancreatic islet cells stimulated to divide in vitro. J Endocrinol 2000; 166:103-109.
61. Parkinson EK, Munro J, Steeghs K et al. Replicative senescence as a barrier to human cancer. Biochem Soc Trans 2000; 28:226-233.
62. Shall S, Stein WD. A mortalization theory for the control of the cell proliferation and for the origin of immortal cell lines. J Theor Biol 1979; 76:219-231.
63. Yuan H, Kaneko T, Matsuo M. Relevance of oxidative stress to the limited replicative capacity of cultured human diploid cells: The limit of cumulative population doublings increases under low concentrations of oxygen and decreases in response to aminotriazole. Mech Ageing Dev 1995; 81:159-168.
64. von Zglinicki T, Saretzki G, Docke W et al. Mild hyperoxia shortens telomeres and inhibits proliferation of fibroblasts: a model for senescence? Exp Cell Res 1995; 220:186-193.
65. Mouton RE, Venable ME. Ceramide induces expression of the senescence histochemical marker, beta-galactosidase, in human fibroblasts. Mech Ageing Dev 2000; 113:169-181.
66. Kim H, You S, Farris J et al. Expression profiles of p53-, p16(INK4a)-, and telomere-regulating genes in replicative senescent primary human, mouse, and chicken fibroblast cells. Exp Cell Res 2002; 272:199-208.
67. Russo I, Silver AR, Cuthbert AP et al. A telomereindependent senescence mechanism is the sole barrier to Syrian hamster cell immortalization Oncogene 1998; 17:3417-3426.
68. Sherr CJ, DePinho RA. Cellular senescence: Mitotic clock or culture shock? Cell 2000; 102:407-410.
69. Li Y, Yan Q, Wolf NS. Long-term calorie restriction delays age-related decline in proliferation capacity of murine lens epithelial cells in vitro and in vivo. Invest Ophthalmol Vis Sci 1997; 38:100-108.
70. Karatza C, Stein WD, Shall S. Kinetics of in vitro ageing of mouse embryo fibroblasts. J Cell Sci 1984; 65:163-175.
71. Karatza C, Shall S. The reproductive potential of normal mouse embryo fibroblasts during culture in vitro. J Cell Sci 1984; 66:401-409.
72. Tollefsbol TO, Cohen HJ. Werner's syndrome: An underdiagnosed disorder ressembling premature aging. Age 1984; 7:75-88.
73. Faragher RG, Kill IR, Hunter JA et al. The gene responsible for Werner syndrome may be a cell division "counting" gene. Proc Natl Acad Sci USA 1993; 90:12030-12034.
74. Kill IR, Faragher RG, Lawrence K et al. The expression of proliferationdependent antigens during the lifespan of normal and progeroid human fibroblasts in culture. J Cell Sci 1994; 107:571-579.
75. Fukuchi K, Martin GM, Monnat RJ Jr. Mutator phenotype of Werner syndrome is characterized by extensive deletions. Proc Natl Acad Sci USA 1989; 86:5893-5897.
76. Lebel M, Leder P. A deletion within the murine Werner syndrome helicase induces sensitivity to inhibitors of topoisomerase and loss of cellular proliferative capacity. Proc Natl Acad Sci USA 1998; 95:13097-130102.
77. Cheng RZ, Murano S, Kurz B et al. Homologous recombination is elevated in some Werner-like syndromes but not during normal in vitro or in vivo senescence of mammalian cells. Mutat Res 1990; 237:259-269.
78. Runger TM, Bauer C, Dekant B et al. Hypermutable ligation of plasmid DNA ends in cells from patients with Werner syndrome. J Invest Dermatol 1994; 102:45-48.
79. Wyllie FS, Jones CJ, Skinner JW et al. Telomerase prevents the accelerated cell ageing of Werner syndrome fibroblasts. Nat Genet 2000; 24:16-17.
80. Schulz VP, Zakian VA, Ogburn CE et al. Accelerated loss of telomeric repeats may not explain accelerated replicative decline of Werner syndrome cells. Hum Genet 1996; 97:750-754.
81. Miller RA. Telomere diminution as a cause of immune failure in old age: an unfashionable demurral. Biochem Soc Trans 2000; 28: 241-245.
82. Goto M, Tanimoto K, Miyamoto T. Immunological aspects of the Werner's syndrome: An analysis of 17 patients. Adv Exp Biol Med 1985; 190:263-284.
83. James SE, Faragher RG, Burke JF et al. Werner's syndrome T lymphocytes display a normal in vitro life-span. Mech Ageing Dev 2000; 121:139-149.
84. Fukuchi K, Tanaka K, Kumahara Y et al. Increased frequency of 6-thioguanine-resistant peripheral blood lymphocytes in Werner syndrome patients. Hum Genet 1990; 84:249-252.
85. Ostler EL, Wallis CV, Aboalchamat B et al. Telomerase and the cellular lifespan: Implications for the aging process. J Pediatr Endocrinol Metab 2000; 13:1467-1476.

86. Ostler EL, Wallis CV, Sheerin AN et al. A model for the phenotypic presentation of Werner's syndrome. Exp Gerontol 2002; 37:285-292.
87. Johnson FB, Marciniak RA, McVey M et al. The Saccharomyces cerevisiae WRN homolog Sgs1p participates in telomere maintenance in cells lacking telomerase EMBO J 2001; 20:905-913.
88. Wei W, Hemmer RM, Sedivy JM. Role of p14(ARF) in replicative and induced senescence of human fibroblasts. Mol Cell Biol 2001; 21:6748-6757.
89. Kill IR. The cell and molecular biology of cellular ageing. D Phil Thesis 1989; University of Sussex, UK.
90. Martin GM, Sprague CA, Epstein CJ. Replicative lifespan of cultivated human cells. Effects of donor's age, tissue and genotype. Lab Invest 1970; 73:3584-3588.
91. Schneider EL, Mitsui Y. The relationship between in vitro cellular aging and in vivo human age. Proc Natl Acad Sci USA 1976; 73:3584-3588.
92. Bermach G, Mayer U, Naumann, GOH. Human lens epithelial cells in culture. Exp Eye Res 1991; 52:113-119.
93. Lipman RD, Taylor A. The in vitro replicative potential and cellular morphology of human lens epithelial cells derived from different aged donors. Curr Eye Res 1987; 6:1453-1457.
94. Blake DA, Yu H, Young DL et al. Matrix stimulates the proliferation of human corneal endothelial cells in culture. Invest Ophthmol Vis Sci 1997; 38:1119-1129.
95. Perillo NL, Walford RL, Newman MA et al. Human T lymphocytes posess a limited in vitro lifespan. Exp Gerontol 1987; 24:177-178.
96. Gilchrest BA. Relationship between actinic damage and chronologic aging in keratinocyte cultures in human skin. J Invest Dermatol 1983; 81:184s-189s
97. Rohme, D. Evidence for a relationship between longevity of mammalian species and life spans of normal fibroblasts in vitro and erythrocytes in vivo. Proc Natl Acad Sci USA 1981; 78:5009-5013.
98. Van Hinsburgh VWM. Arteriosclerosis: Impairment of cellular interactions in the arterial wall. Ann NY Acad Sci 1992; 673:321-330.
99. Hoppenreijs VPT, Pels E, Vrensen GFJM et al. Effects of platelet derived growth factor on endothelial wound healing of human corneas. Invest Ophthmol Vis Sci 1994; 35:150-161.
100. Linskens MHK, Feng J, Andrews WH et al. Cataloging altered gene expression in young and senescent cells using enhanced differential display. Nuc Acid Res 1995; 23:3244-3251.
101. Gray MD, Norwood TH. Cellular aging in vitro. Rev Clin Gerontol 1995; 5:369-381.
102. Dimri G, Lee X, Basile G et al. A biomarker that identifies senescent human cells in culture and in aging skin in vivo. Proc Natl Acad Sci USA 1995; 92:9362-9367.
103. Schultze B, Korr H. Cell kinetic studies of different cell types in the developing and adult brain of the rat and mouse: A review. Cell Tissue Kinetics 1981; 14:309-325.
104. Tannock IF, Hayashi S. The proliferation of capillary endothelial cells. Cancer Res 1972; 32:77-82.
105. Hyldahl L. Control of cell proliferation in the human embryonic cornea: An autoradiographic analysis of the effect of growth factors on DNA synthesis in endothelial and stromal cells in organ culture and after explantation in vitro. J Cell Sci 1986; 83:1-21.

Index

A

Aging 1, 2, 5, 6, 9, 12-14, 16, 18, 44, 58, 62, 63, 69, 74, 78-81, 87, 88, 92, 94-96, 103-107, 111, 113-115, 117, 123, 129, 133-135, 137-140, 146, 147
Alopecia 3, 5
Alternative lengthening of telomeres (ALT) 17, 18, 91-93, 141
Anti-topoisomerase I 7
Apoptosis 15, 16, 45, 46, 50, 51, 58, 64, 66, 70-72, 74, 90, 111, 116
Ataxia telangiectasia (AT) 62, 63, 72, 83, 88, 140
Atherosclerosis 3, 5, 7, 9, 12, 111, 115
Autosomal recessive inheritance 2

B

Base excision repair (BER) 45-47, 50, 70, 71, 125
BCR-Abl 55
BLM 12-14, 16-18, 24, 29-32, 34, 37, 45, 46, 49, 52, 56-58, 63, 64, 68, 71, 72, 78-80, 82-86, 90-92, 94, 96, 97, 108, 113
Bloom's syndrome (BS) 12-14, 24, 29, 30, 52, 57, 62, 63, 74, 78, 79, 83, 85, 90, 96, 113
Bubble-containing duplex DNA 22, 29-31, 38

C

c-Abl 52, 55
Cancer 2, 3, 8, 12-14, 44, 50-52, 62-64, 69, 72, 78, 83, 94, 107, 111, 115-118
Cataract 1-3, 5, 9, 12, 14, 63, 111, 115
Cell cycle checkpoint 17, 80, 90
Cockayne's syndrome (CS) 14, 62, 94
Cushingoid appearance 5, 7

D

Diabetes mellitus 3, 5, 12, 63, 111, 115
DNA double strand break 52, 66, 71, 88, 125
DNA polymerase β 45-47
DNA polymerase δ 15, 38, 44-47, 50, 57, 69, 124
DNA recombination 29, 50, 51, 82, 107, 112, 117, 125, 127
DNA repair 12, 14-16, 18, 27, 46, 47, 50, 53, 56, 62, 69, 70, 72, 73, 78, 79, 81, 82, 86-88, 90, 91, 94, 96, 111, 117, 125, 127
DNA replication 12, 15, 16, 44-47, 50, 55-58, 62, 63, 69-72, 78, 79, 82, 83, 85, 87, 88, 90-93, 95-97, 107, 109-112, 117, 123-125, 128, 129, 139
DNA-dependent protein kinase (DNA-PK) 16, 17, 37, 45, 52, 54, 55, 70, 72, 83, 89, 90, 125-128
DNA-PK$_{cs}$ 44-46, 52, 54, 57, 125-128
Double strand break 45, 52, 54, 56, 57, 87, 88, 90, 117, 125-127, 140
Double strand break repair 66, 70, 125

E

EXO-1 44-47
  see also Human exonuclease-1
Extrachromosomal rDNA circle (ERC) 94, 95

F

Fanconi anaemia (FA) 63
FEN-1 15, 38, 44-49, 56, 124, 125, 127
Fibroblast 2, 3, 16, 17, 29, 36, 51, 63, 64, 79, 93, 107, 109, 111, 113, 115, 127, 128, 133-137, 139, 140, 142-147

Fluorescence in situ hybridisation (FISH) 63, 64
FOB1 95
Focus forming activity-1 (FFA-1) 13, 24, 47, 57, 69, 108-111
Forked DNA 27, 29, 31, 35, 80, 81

G

Genetic instability 9, 117, 118, 128
Genomic instability 13, 14, 16, 24, 30, 44, 50-52, 58, 64, 72, 74, 78, 79, 83, 93, 94, 96, 107, 112, 115-117, 124, 125, 127-129, 146
Gout 5
Graves' disease 5-7

H

Helicase 1, 2, 9, 12-18, 22-38, 44, 45, 47, 49-52, 54, 55, 57, 58, 62-74, 78-97, 107-113, 115-117, 123-127, 133, 147
HEX3/SLX8 86
Holliday junction 16, 18, 27, 29, 30, 37, 38, 45, 50, 51, 53, 55, 58, 68, 80, 83, 91, 92, 127
Homologous recombination repair 57, 71

hTERT 93, 142, 146
Human exonuclease-1 47
 see also EXO-1
Human replication protein A 45, 67
 see also replication protein A (RPA)
Hyperlipidemia 3, 5, 6
Hypogonadism 3, 5, 111
Hypoxanthine-guanine phosphoribosyl-transferase (HPRT) 64, 112, 146

I

Ionizing radiation (IR) 54, 63, 64, 70, 86, 125, 126

K

Ku 16, 17, 33, 37, 38, 44, 46, 52-54, 56, 57, 70, 83, 90, 117, 125-128
Ku complex 52, 54, 56, 57, 117

L

Longevity 94, 96

M

Meiosis 29, 79, 82, 84-87, 96, 97
Melanoma 3
Mismatch repair (MMR) protein 81, 83
MMS4/MUS81 86, 87

N

NK cell 5, 6
Non-homologous end joining (NHEJ) 44-46, 52, 54, 57, 58, 88, 93, 94, 125-128
Nucleotide excision repair (NER) 63, 64, 86, 87, 94

O

Osteoporosis 3, 5, 12, 63, 111, 115

P

p21 14, 116, 140
p53 14-17, 33, 37, 38, 44-46, 50-52, 55, 56, 70-72, 89, 90, 115-117, 127, 140, 141, 143, 145
p97/VCP 45, 46, 55
Poly(ADP-Ribosyl) Polymerase-1 (PARP-1) 44, 46, 50
Polymerase δ (Pol δ) 15, 16, 28, 38, 44-48, 50, 57, 69, 124, 125
Progeria adultorum 1
Progeroid syndrome 1, 9, 63, 134, 137
Proliferating cell nuclear antigen (PCNA) 15, 16, 38, 44-50, 55, 57, 67, 69-72, 84, 90, 91, 110, 112, 115, 116, 124, 125, 146

R

Rad16 81, 86
RAD51 15, 16, 18, 44-46, 49, 51, 53, 54, 56-58, 68, 70-72, 74, 84, 86-88, 91-94, 127
RAD52 18, 44-46, 51, 52, 54, 56, 57, 92
Recombination 12, 14, 15, 17, 18, 26, 27, 29-31, 34, 37, 38, 44-46, 49-52, 54, 56-58, 62-66, 68-72, 78-80, 82-88, 91-94, 96, 97, 107, 109, 111, 112, 116, 117, 123-125, 127-129, 140
RecQ 2, 12-14, 16-18, 22-25, 28, 30, 32, 33, 37, 45, 47-49, 52, 56-58, 63, 65, 67, 68, 70-72, 74, 78-86, 88-94, 97, 107, 108, 111, 113, 124, 127, 128, 133
RecQ helicase 12-14, 16-18, 23-25, 28, 32, 37, 45, 47, 49, 52, 57, 58, 65, 68, 70-72, 74, 78-86, 88-93, 97, 108, 133
RecQ3 1, 2, 9
RecQ4 12-14, 78, 80
Replication fork 15, 29, 31, 45, 48-51, 53-58, 68-72, 78, 83, 87, 88, 90, 91, 93, 95, 107, 110, 112, 117, 124, 125, 127, 128, 133, 147
Replication protein A (RPA) 15, 16, 25, 38, 44-47, 49, 56, 57, 63, 67-72, 79, 84, 91, 97, 110, 123-125, 127
Replicative senescence 17, 18, 70, 72, 94, 128, 133-140, 142, 145-148
Rothmund-Thomson syndrome (RTS) 12-14, 24, 94

S

S phase 44, 51, 63, 66, 69-72, 78, 79, 82, 86-91, 112, 113, 116, 123, 138, 140-142, 144
Sarcoma 9
Schizophrenia 3, 5, 9
Scl 70 7
Scleroderma 3, 5

SGS1 (Sgs1) 13, 16, 18, 24, 28, 29, 50, 55, 56, 64, 65, 68, 69, 71, 72, 78-97, 108, 127, 128
Sister chromatid exchange (SCE) 50, 52, 57, 63, 74
SLX1/4 86
SRS2 87, 88, 95
SUMO-1 17, 45, 46, 55, 84, 90
Systemic lupus erythematosus (SLE) 5, 7

T

Telomerase 17, 18, 29, 72, 79, 80, 91-93, 128, 141-143, 146
Telomere 14, 16-18, 26, 28, 29, 34, 45, 68, 72, 78-83, 85, 91-96, 117, 125, 128, 140-147
Telomere maintenance 16-18, 68, 72, 78-80, 82, 92, 93, 128
Tetraplex DNA 26-29, 32, 38
TOP1 85, 86
TOP2 85
TOP3 85, 87, 90
Topoisomerase I (Topo I) 7, 15, 44-46, 49, 55-57, 64, 66, 85, 86, 110, 112, 113, 117, 118, 124, 125
Topoisomerase II (Topo II) 45, 46, 56, 65, 66, 85
Transcription 12, 14, 16, 23, 24, 30, 35, 36, 62, 63, 68, 69, 71, 72, 81, 82, 84, 88, 94, 109, 111, 112, 115, 127, 128, 139-142
Trichothiodystrophy (TTD) 14
Triple-helical DNA 22, 30
Triplex DNA 29, 30, 68

U

UBC9 (Ubc9) 16, 17, 45, 46, 55, 84, 90
Ubiquitination 55

W

Werner helicase interacting protein
  (WHIP) 16, 44-46, 49, 84, 92, 16
WHIP/WRNIP1 44, 45, 49
WRN (Wrn) 1-3, 9, 12-18, 22-38,
  44-58, 62-74, 78-80, 82-87, 89-94,
  96, 97, 107-118, 123-129, 133,
  144-147
WRN exonuclease 15-17, 33, 34, 51,
  52, 54, 58, 67, 70, 72, 108, 125
WRN (Wrn) helicase 13, 15, 16, 18,
  22-27, 31, 32, 35-38, 51, 54, 57,
  58, 62, 64, 66-70, 72, 73, 93, 108,
  112, 113, 115, 116, 123, 125-127,
  133

X

Xeroderma pigmentosum (XP) 14, 62,
  63, 64